TOMORROW'S MATERIALS
Second Edition

TOMORROW'S MATERIALS

Second Edition

Ken Easterling

Formerly of the
Department of Engineering Materials
University of Luleå
S-95187 Luleå
Sweden

THE INSTITUTE OF MATERIALS

Book 493

Published by
The Institute of Materials
1 Carlton House Terrace
London SW1Y 5DB

This edition © 1990 The Institute of Metals; reprinted 1993
(now The Institute of Materials)
First edition published 1988; reprinted 1989 (twice)

All rights reserved

British Library Cataloguing in Publication Data
Easterling, Kenneth E. (Kenneth Edwin), 1933-
Tomorrow's materials - 2nd ed.
1. Materials science
I. Title
620.11

ISBN 0-901462-83-7

Cover design: Jenny Liddle

Text processing by P *i* c A Publishing Services,
Abingdon, Oxon.

Printed and bound in Great Britain by
Bourne Press, Bournemouth

To Edward, William, Elizabeth-Jane
and other citizens of tomorrow's world

Contents

Foreword ix
Preface to first edition xi
Preface to second edition xiii

PART I FUNDAMENTALS   **1**

Introduction 2
Crystalline and amorphous materials 5
Solidification from the melt 15
Composite and cellular materials 21
Why metal bends, rubber stretches, and glass breaks 25
Materials selection 30
Characterising materials 36
Further reading 41

PART II APPLICATIONS   **43**

Structural materials 46
Lightweight materials 55
Wear- and heat-resistant materials 71
Optical materials 91
Electronic and magnetic materials 100
Sporting materials 119
Further reading 140

viii   *Tomorrow's Materials*

PART III TOWARDS TOMORROW'S WORLD   **141**

Planet Earth: Ecosphere or dustbin? 142
The strategic elements 146
Energy and population 148
Recycling: the materials merry-go-round 154
Biodegradeable polymers: drops in a plastic bucket 158
The crunch: congestion or consciousness? 160
Further reading 163

Glossary 164

Index 171

Foreword

As every schoolboy knows, the ages in which man has lived and progressed are named for the materials he used: stone, bronze, iron. And when he died, the materials he valued were buried with him: Tutankamun with shards of coloured glass in his stone sarcophagus, Agamemnon with his bronze sword and mask of gold, each representing the high technology of his day.

If they had lived and died today, what would they have taken with them? Artefacts of microalloyed steel, of polyether-ethyl-ketone, of yttria-toughened zirconia, of kevlar-reinforced epoxy. This is not the age of one material; it is the age of an immense range of materials. There has never been an era in which the evolution of materials was faster and the range of their properties more varied. The menu of materials available to the engineer has expanded so rapidly that designers who left college thirty years ago can be forgiven for not knowing that half of them exist. But not-to-know is, for the designer, to risk disaster. Innovative design, often, means the imaginative exploitation of the properties offered by new materials. And for the man in the street, the schoolboy even, not-to-know is to miss one of the great developments of our age: the age of advanced materials.

This concise and readable book aims to introduce the reader to the recent developments in the science of materials, and to alert him or her to the possibilities that the future holds. It is written, first, for high school leavers who are contemplating a career in science or technology; often their schooling has taught them very little about the role of materials and the opportunities that materials-related professions offer. And it is written, second, for people working in the materials or related industries who would like an overview of the potential offered by new materials. Professor Easterling has written a popular introduction to the fundamentals and likely developments of materials, avoiding the use of equations or detailed theory, and emphasising the far reaching influence of new materials on modern technology.

This book aims to whet the reader's appetite, to draw him or her on, to stimulate an interest in this broad and rapidly developing field. It leaves more complex and detailed explanations to other weightier, more specialised volumes. It is a book which makes the reader wish for more. And that was exactly the author's aim.

Professor Mike Ashby
Department of Engineering
University of Cambridge

Preface to first edition

I have written this book primarily for two groups of people. In the first group are high-school students and school leavers contemplating a career in science or technology. It is only too common to hear from them that, while they know more or less what mathematicians, physicists, or even engineers are supposed to do, they have no idea about that mysterious band who call themselves materials scientists. In the second group are those working in the materials or related industries who need a brief overview of recent advances in new materials. For both groups, I have attempted to write a popular introduction to the fundamentals and likely developments of materials, avoiding the use of detailed theory. The description is far from comprehensive. Materials experts who read this book will find that a number of potentially important materials have been omitted. This is partly due to the limitations of a short and, I hope, 'readable' text. It must be admitted, however, that the materials discussed are those that particularly excite me and seem most likely to become prominent in the near future. There are no doubt others, unmentioned and perhaps undiscovered, which may well surpass them.

The text is divided into two main parts. The first, dealing with the fundamentals of materials science, gives the non-specialist an introduction to the subject and presents a novel way of classifying the various types of materials. The second part deals with applications of advanced materials and is divided into main areas of application rather than types of material. The theme developed in Part I, of avoiding the traditional classification of materials into metals, ceramics and polymers, is thus maintained in Part II, since I feel this better serves the needs of engineering and materials designers.

I should like to thank many people for helping me in the preparation of the manuscript. Of these I must mention my secretary,

xii *Tomorrow's Materials*

Sue Tuohy, for her extremely patient typing and word-processing, and Max Renner and his drawing office staff at the University of New South Wales for their excellent work on the figures. Thanks are also due to Roland Lindfors of the University of Luleå for some of the photography. I am particularly grateful to Bo Bengtsson and Hans Bentilsson, both of the University of Luleå, to Bruce Harris, Alan Crosky, and Chris Sorrell of the University of New South Wales, and to Charles Tonkin, a pupil of Sydney Grammar School, for critically reading the manuscript and suggesting a number of useful amendments and additions.

Ken Easterling
Sydney, May 1987

Preface to the Second Edition

The second edition of this book is a somewhat different animal from the first. To begin with it is more hefty. In fact, it is over 50% thicker than the former, with new sections on Materials Characterisation (Part I) and Sporting Materials (Part II). It also contains a new Part III, dealing with materials resources, recycling and the environment. I've called this section: 'Towards Tomorrow's World' and I try to emphasise the need for more consciousness and less consumerism if we (you) are to really enjoy the benefits of tomorrow's materials.

The book contains a few updatings, although in general its contents have withstood the passage of time well. For completion, I have added a few equations and included a comprehensive index. Finally, readers may be pleased to note that there are now the requisite *seven* wonders of the materials world.

I should like to acknowledge several people (not mentioned in the first preface) who have helped me, both by their constructive criticism of the first edition, and by their contributions to the second. These include: Dr. Jack Harris (Berkeley Nuclear Laboratories), Mr. Lennart Wallstrom (University of Lulea), Mr. Peter Bonfield (University of Bath), Professor John Steeds (University of Bristol), Dr. John Ion (University of Lappeenranta), Dr. Hugh Casey (Los Alamos National Laboratory). I should also like to thank my daughter, Dr. Elizabeth Easterling, for her chiropractic advice in the Sporting Materials Section.

Ken Easterling,
Bristol, January 1990

PART I
FUNDAMENTALS

Four basic families of materials are considered in this first part: amorphous materials, in which atoms are arranged randomly; crystalline materials, in which atoms are stacked up in almost perfect order; composites, consisting of two or more different materials; and cellular materials, consisting of orderly stacks of hollow symmetrical cells. Each of these families includes metals, polymers and ceramics in various forms and proportions. The design of materials depends on our ability to manipulate these forms and proportions, in many cases by modifying the atomic and electron configurations through alloying and heat treatment.

Tomorrow's Materials

Introduction

ET - extraterrestrial? No, it's short for ethylenedithio tetrathiaful-valene, a completely new type of metal which is superconducting at the unusually 'high' temperature of around 270°C below the freezing point of water! Actually, much higher superconducting temperatures have now been achieved, in a new breed of oxide ceramics. These are just two of the new materials that we may meet in tomorrow's world - the world of the twenty-first century. It will not be such a different world to the one we live in today, perhaps, though it will be sufficiently changed to be noticeable.

You'll be driving a somewhat lighter car than your current one, partly because plastics and cellular composites will have replaced about 30% of the steel in its construction, and partly because much of the engine will consist of lightweight nitrogen ceramics which allow it to run at a higher temperature. These changes should please you, because the fuel consumption will be around half of what it is today. You'll drive to work over a fibre-reinforced 'flexible' concrete bridge of a new lightweight tubular construction. The thin concrete sections, manufactured at the factory, will be glued together on-site using a new super-strong polymer glue. The telephone and the electrical and electronic functions in the car will be controlled by a tiny computer based on gallium arsenide, or maybe even plastic (polyacetylene) chips.

Besides new materials, you may notice quite new uses of conventional materials. For example, if the car you are travelling in happens to be a hearse, and this is your final journey, the chances are that you will be lying in a lightweight, foamed-polystyrene, moulded plastic coffin. (Currently there are problems with this particular product because the material cannot yet be made to be biodegradable; however, they do burn well and without too much smoke.)

These predictions are not the result of gazing into a crystal ball, but are based on an appraisal of current research in leading materials laboratories around the world. Since some ten to fifteen years usually pass between the beginning of application-oriented

Fundamentals

research and the production line, tomorrow's materials, based on the most promising of today's research, may be almost commonplace by the turn of the century.

Of course, in discussing the likely materials of tomorrow's world it is not enough to consider current research. In a dynamic world of changing political frontiers, shrinking distances, widening information networks, and growing environmental concern, attitudes to the production and use of materials will also change, probably radically. All current trends in materials research point to a much more efficient use of materials in the future. For materials which are becoming scarce this approach is obvious, and research must aim to find alternative materials to replace the old ones. But this is only a small part of the picture. Materials of the future will be produced more efficiently than ever before, with emphasis on fewer operations and much less wastage in producing the final shape. Recycling old materials is already an area of growing importance, and in this respect the non-biodegradable coffin and many other synthetic polymers will be the objects of considerable research. The Italian Government has decreed that only degradable plastic bags will be available for shoppers after the year 1992. In other words, tomorrow's materials will provide the basis of a manufacturing technology which is more energy-conscious and environmentally responsible than today's, in terms of both material production and application.

You might think that we already have quite enough materials to see us into the next century. A look at manufacturers' catalogues reveals tens of thousands of materials from which to choose. In spite of this, tomorrow's engineering designer will certainly have a great deal more from which to choose than his predecessors of today. For example, although there are currently about 15 000 different plastics available, it is predicted by scientists that this number will double by the year 2 000. This is not to imply, of course, that all today's materials will continue to be manufactured tomorrow. On the contrary, changing attitudes to use and production will result in better, more sophisticated materials, designed to widen their application and provide more economical and lighter structures. Furthermore, metallurgists

3

Tomorrow's Materials

and ceramists are as active as their polymer colleagues in developing new products, and competition between these industries may well sharpen in the future.

The new century will witness completely new types of materials. We are now getting some spectacular insights into tomorrow's materials from the natural world, from investigations of cross-sections of dragonflies' wings, the composition of spiders' legs, the mechanics of holly leaves, the microstructure of seashells and coral, and the 'super-glue' used by mussels. The purpose of this work is to find how to apply the techniques of Nature to the design of tomorrow's lightweight cellular and composite structures. Figure 1 shows part of the cross-section of the leaf of a lily. The results of studies like this may lead to better designs for sandwich panels (these are honeycomb structures glued between hard outer sheets).

Against a background of changing attitudes, closer interaction between tomorrow's materials scientists and engineering designers is of paramount importance. This is necessary not only to enable designers to take advantage of the exciting new materials becoming available, but also to provide a basis for developing even better materials and technologies in the future. Even now, materials scientists know so much about what gives materials their various thermal, optical, electrical, and mechanical properties, that they can in effect design materials to fit a particular application in the same way that the mechanical engineer designs the shape of struts in a bridge or panels in a spacecraft.

Before getting down to the finer details of tomorrow's materials, it is useful for the non-specialist to consider briefly what it is that gives materials the properties they actually possess. For example, when is a metal not a metal? When do polymers behave more like metals? What's the difference between a plastic and a ceramic? We shall see that while the traditional division of materials into metals, polymers, and ceramics has been useful in the past, in the modern, wider approach to the structural properties and applications of materials it is better to consider materials under more general headings. When a designer comes to select a material for a particular application, all possible materials will

4

Fundamentals

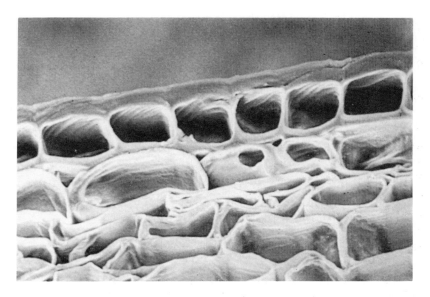

1 This electron micrograph of the cross-section of the leaf of a lily may give engineers clues to the design of sandwich panels   (Magnification x 700)

then begin on a more or less equal footing, irrespective of their species. We shall see that a suitable division of materials can be made under the four headings crystalline, amorphous, composite and cellular. We begin with a discussion of crystalline and amorphous materials.

Crystalline and amorphous materials
Order versus chaos in the world of materials

The physicist and Nobel Laureate Sir Nevill Mott has suggested that all materials can be said to be either amorphous or crystalline. The atoms of crystalline materials are stacked up in perfect ordered symmetrical arrangements, while in amorphous materials agglomerations of atoms have little or no order or symmetry. A two-dimensional analogue would be the comparison of a regiment of soldiers on the parade ground with a mass of people in a crowded railway station. The division is not absolute, although in some ways this depends how the term 'amorphous' is defined. Many synthetic polymers, for instance, have a structure which contains a mixture of amorphous and crystalline material, depending on how the individual polymer chains are arranged with respect to their neighbours. As shown in Figure 2, various regular or repeating arrangements can be achieved such that a sort of regimented structure is developed within a matrix of randomly arranged chains. It should not be forgotten that the chains themselves are far from random arrangements of atoms, but are highly regular and repeating arrays of molecules, as shown schematically in Figure 3. In spite of this, as a general rule polymers are regarded as amorphous materials.

2 The atomic chain structure of a polymer, showing random (amorphous) and ordered (crystalline) regions

With metals the distinction between crystalline and amorphous arrangements of atoms is more apparent, one being completely regular and the other quite random, as illustrated in Figure 4. Either of these atomic structures can be obtained, the final product depending mainly upon the rate of cooling from the liquid state. Very rapid cooling (by about a million degrees per second) can effectively 'freeze-in' the random arrangement of

Fundamentals

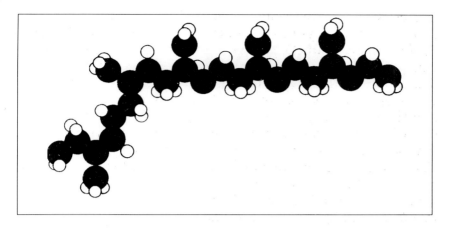

3 A single polymer chain, showing the arrangement of carbon atoms (large black circles) and hydrogen atoms (small white circles)

atoms in the liquid, and such metals are called 'glassy' because of their similarity to the amorphous atomic structure of ordinary glass. However, as seen from Figure 5, the atomic structure of glass is not exactly like that of an amorphous metal; there has to be some short-range molecular order of the atoms so as to maintain the strict silicon/oxygen molecular structure, or 'stoichiometry', of the glass. Nevertheless, like polymers these materials are still usually classed as amorphous.

Crystalline materials are different from amorphous materials in that they can undergo plastic deformation. Plastic deformation in metals, for example, is made possible by the presence of defects called dislocations, which in effect allow atoms in one crystal plane to move relative to an adjacent plane, as illustrated in Figure 6. The left-hand side of Figure 6 shows how a dislocation moves by distorting the surrounding crystal lattice. At the core of the dislocation, atoms move in discrete little jumps as illustrated at the right of Fig.6. In an electron microscope, in which magnifications be achieved of several hundred thousand times, dislocations can be actually observed as thin wavy lines. These can sometimes be seen moving within the thin crystal in a jerky fashion as impurity atoms hinder the dislocation's glide in places. The movement of a large number of dislocations together brings about

Tomorrow's Materials

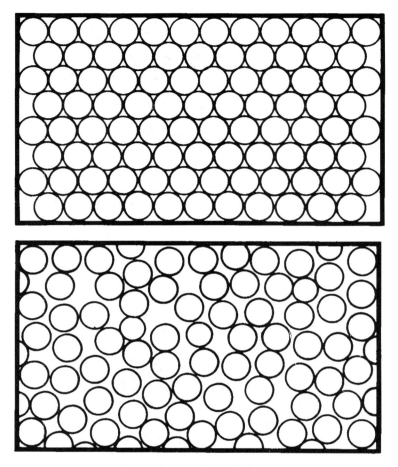

4 Comparison between (a) the perfect atomic regularity of a crystalline metal and (b) the disordered atomic structure of an amorphous metal

plastic (irreversible) deformation in a crystal; this process is illustrated schematically in Figure 7. It can be shown that the initiation and growth of cracks is controlled by dislocation movement. In effect, a crack can be likened to a concentrated group of dislocations occurring, for example, at a crystal boundary or an inclusion (a small impurity).

Fundamentals

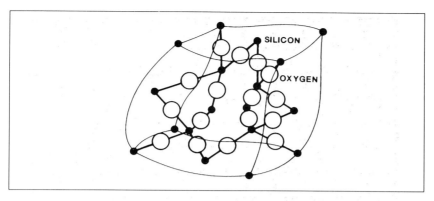

5 The random three-dimensional network of a silica glass, here likened to the distorted cubic lattice of crystalline silica

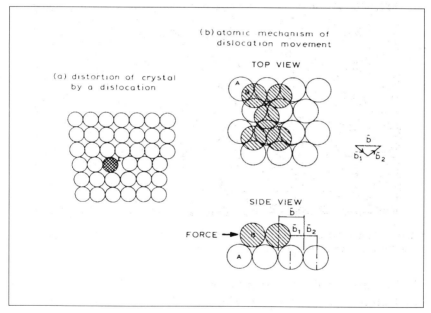

6 Dislocations allow the crystal segments of a metal to slide over one another by the movement of atoms. The vector diagram represents the individual movements of atoms in terms of 'partial dislocations'. The sum of these 'partials' is called the Burgers vector, after the person who conceived the idea (see box on p.10). The movement of a dislocation actually severly strains the surrounding crystal as illustrated on the left side of the figure. As the shaded atom moves right, the dislocation moves left.

Tomorrow's Materials

Forces to move Dislocations

If a piece of metal is loaded such that it starts to bend, in effect lots of dislocations within the metal have been made to move. The movement of each dislocation can be characterised in terms of the movements of atoms at its core, such that the applied force overcomes the friction of dislocation movement through the crystal. The equation describing this is thus:

$F = \mu b$

Where F is the force, μ is the frictional stress within the glide plane, and b is the Burger's vector of the dislocation. Since μ can be measured by experiments on single crystals of the metal, and b is simply (typically) the atom size of the metal (which is known), F can be estimated fairly accurately. It is found in most cases to be very much smaller than that needed to fracture the metal, which is why pure metals can be bent so easily.

All crystalline materials, particularly metals, contain millions of dislocations per cubic millimetre. Indeed, it is due to their presence that metals are so easily deformed. While it is not so easy to create new dislocations, existing ones move relatively easily under an applied external stress (see box). For example, in bending a spoon, great activity occurs as dislocations are created and

Fundamentals

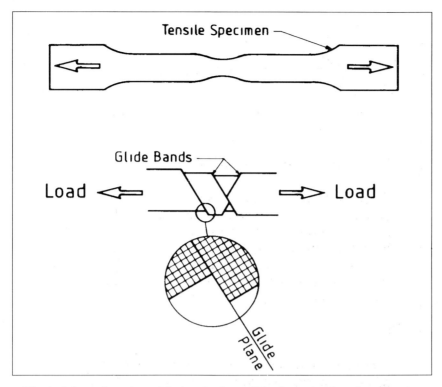

7 Plastic deformation of a metal, showing how glide planes develop along the direction of highest shear stress in a tensile specimen, a testpiece stretched along its axis

rearrange at velocities approaching the speed of sound, to acomodate the spoon's new shape.

While dislocations in this sense do not exist in amorphous materials, they do occur in crystalline ceramics. However, ceramics have a more complicated molecular arrangement of atoms than is found in metals, and this makes dislocation movements of the sort illustrated in Figure 7 very complicated. An example of the crystal structure of a typical ceramic material is shown in Figure 8, which illustrates the molecular relationship between oxygen and silicon atoms. When dislocations move through crystalline ceramics, the molecular arrangement has somehow to be maintained. Coupled

Tomorrow's Materials

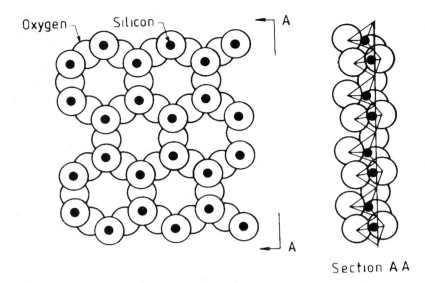

8 Ceramics are crystalline compounds with ordered atomic arrangements, often between metallic and non-metallic elements such as silicon or aluminium and oxygen or nitrogen. The bonding between these atoms is usually ionic or covalent, making them more directional and rigid than metallic bonds

with this, as we shall see below, the bonds between atoms in ceramics are more rigid than in metals, which makes plastic deformation a difficult process. Ceramics are therefore usually rather brittle. Indeed, it might be said that metals and ceramics represent two extremes of crystalline behaviour, from good ductility and high toughness (metals) to poor ductility and low toughness (ceramics). However, most ceramics are mechanically much harder than metals and usually have higher melting points, because of the powerful (oriented) molecular bonding in these materials.

An essential feature of atomic bonding in any material is the way in which electrons are shared between atoms. Indeed, a generally accepted advantage of the traditional classification of materials into metals, polymers, and ceramics is that the properties of each class of material can be related to the atomic bonding and electron configurations (see Figure 9). In other words, the atomic and electron configurations largely determine the charac-

Fundamentals

teristics of each type of material. Thus, for example, metals are typically good conductors of electricity and heat because of their regular atomic arrangement and free valence electrons. Metals are also plastically deformable, a state in which metals can be squeezed like Plasticene to change their shape, as in rolling or extrusion operations. This is possible because of the nondirectional nature of the atomic bonding in metals, which in effect allows atoms and planes to slide over one another when subjected to sufficiently high mechanical loading (see Figure 7). Ordinary plate glass, a ceramic, is transparent, whereas metals are not. This is because glass is amorphous - the atoms are disordered, and there are no reflecting 'surfaces' such as the grain boundaries which make metals opaque. In addition, the bound electron shell configurations in this material allow light to pass through without significant photon scattering occurring. Glass is brittle because of its rigid atomic bonds and non-crystalline structure.

Many other ceramics are crystalline, yet are poor conductors as they lack free electrons. However, because of their crystallinity and finegrained microstructure these materials are usually opaque. Also, most ceramics, like glass, have limited plasticity (except at very high temperatures) because of their highly rigid and directional atomic bonding. Most polymers are poor conductors (they lack free electrons) and cannot be plastically deformed in the same way as metals, again because they consist of rigid chains of atoms. However, some polymers (like rubber) can be deformed by straightening out their component chains. Other polymers in this category are polyesters, polyamides, and polycarbonates. Brittle polymers include Perspex (or Plexiglas) and polystyrene. Many polymers are transparent by virtue of their amorphous atomic structure.

In these examples, 'most' members of a class of materials have been described as 'usually' having a particular property. It is worth pointing out that current research is already producing many interesting exceptions. 'Glassy' metals have already been mentioned. Their random but relatively close-packed atomic structure provides them with unique magnetic properties. They may also be more easily superconducting, a state in which resis-

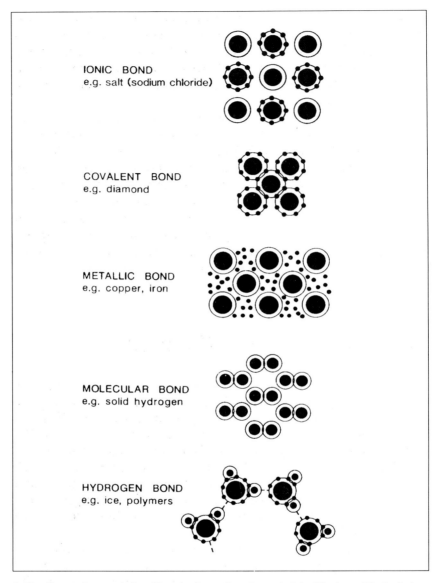

9 The five main atomic bonding configurations in materials. The large black circles represent atoms, and the small ones electrons. The rings surrounding atoms represent electron orbitals. Note that the very strong oriented covalent bond consists of shared electrons. In metals, atoms are surrounded by an 'atmosphere' of free (valence) electrons

tance to electron flow is reduced virtually to zero. However, they show little or no plasticity, and in this respect do not behave as typical metals. At the other extreme, Japanese scientists have recently developed a superplastic ceramic. Based on fine silicon nitride and carbide grains in a "glue" of oxide, this material is more plastic than most metals.

An interesting recent development concerns electrical conductors made of polymers. These are made by 'doping' certain polymers with salts and iodides, which makes the electron configuration of the polymer similar to that of semiconductors. Such polymers, dubbed 'polymer metals', exhibit 100 billion times the normal conductivity of typical polymers! It has even been demonstrated that ceramics, traditionally considered to be resistors, can also be made semiconducting and even superconducting by suitable alloying. In other words, polymers and ceramics can be made to behave very much like metals in some respects.

Solidification from the melt

When a liquid is cooled below its freezing point, whether it forms into a random amorphous solid or an ordered crystalline one depends on the rate at which the liquid is cooled. Solidification usually results in an abrupt decrease in volume, because the mobility of atoms in the solid state is lower than in the liquid. (A well-known exception is the solidification of water to ice, which results in an expansion.) The volume change during the transition from a liquid to a solid in amorphous materials occurs gradually, since there is no change in the arrangement of these atoms as in crystalline solidification.

In crystalline solidification, provided the liquid is cooled sufficiently slowly a temperature is reached at which small, close-packed regular groups of atoms, or crystal 'nuclei', form. On lowering the temperature still further, these nuclei grow and finally the crystals merge together as illustrated in Figure 10. Since the original groups of atoms or crystal nuclei are not necessarily aligned with one another at the beginning of solidification (see Figure 10 (a)), the resulting solid crystal is not a single crystal but an agglomeration of several crystals - a 'polycrystal'. Each crystal is separated from its neighbours by crystal or grain boundaries (see Figure 10 (c) and (d)) which act as planar defects, breaking up the continuity of perfect atomic stacking of the solid as a whole. Grain boundaries are actually very important to the properties of crystalline materials, for this reason. In polycrystals, plastic deformation of the type illustrated in Figure 7 has to occur on different planes (and directions) in each crystal. Therefore, in the plastic deformation of polycrystals the whole solid has to deform in such a way that cracking does not take place between individual grains.

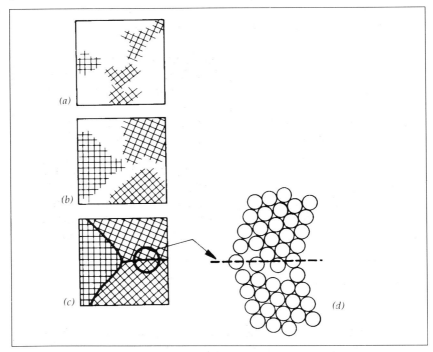

10 How polycrystalline metals form by solidification from the melt, starting with (a) branched structures called dendrites, which (b) continue to grow until (c) they meet other growing crystals to form grain boundaries (d)

Fundamentals

Provided these crystal boundaries are reasonably free of impurities, polycrystals are stronger, or more resistant to plastic deformation, than single crystals. This is an important consideration in the manufacture of practical high-strength metals and alloys; in fact, it can be shown that the strength of a metal increases as the grain size decreases (see box).

The Hall-Petch Equation of Strength

Two researchers working in different parts of the world, Dr. E. Hall of Newcastle, New South Wales, Australia and Dr. N. Petch of Newcastle, England, arrived almost simultaneously at an important equation describing the strength of a polycrystalline metal, without being aware of the other's work. They found that the externally applied stress to cause a metal to flow plastically, s_a depends on the (macroscopic) frictional stress to move dislocations in the metal, s_o, and the way dislocations pile up at grain boundaries to cause dislocation movement into the next grain. The latter phenomenon is denoted by a constant, K. The dependence of flow stress on grain size, d, is thus written:

$$s_a = s_o + \frac{K}{\sqrt{d}}$$

This equation is the basis of the design of modern structural metals, in which the yield strength is maximised by making the grain size as small as possible.

Tomorrow's Materials

The nucleation and growth of crystals is a process that requires time. If the liquid is cooled at such a rate that there is no time for ordered crystalline groups of atoms to nucleate and grow, the liquid freezes to a disordered amorphous solid instead. For example, when silica plate glass solidifies the liquid first becomes viscous, rather like thick treacle. In this state atoms no longer have the freedom to change their position relative to their neighbours and form into ordered groups. Thus on further reduction of temperature the silica molecules freeze into the amorphous state of glass, because there is no systematic close-packed stacking of atoms as in crystallisation.

In the formation of plate glass the liquid-to-amorphous solid transition is achieved at quite moderate rates of cooling. In metals a very rapid decrease in temperature from the melt is required if crystallisation is to be avoided. By controlling the rate of cooling from the liquid most materials can be solidified to the glassy state, but not all materials can be made to crystallise. Synthetic polymers and natural organic compounds like alcohol, glycerol and glucose, for example, form an amorphous structure when solidifying, even at a relatively gentle rate of cooling. Whether polymers form purely amorphous (irregular chain arrangements), or mixed amorphous/crystalline structures depends upon the shape of their molecular arrangements, and not upon the cooling rate.

Solidification takes place because of a thermodynamic energy difference between the molten and solid phases. This thermodynamic factor is a rather complex one and takes into account the relative bonding energies of the atoms in the liquid and solid states (see box). It varies enormously from one material to another, which is why some materials can easily be cooled to the glassy state and others cannot. Thus to take two extreme examples, a cooling rate of up to a million degrees per second is needed to produce glassy metals, whereas a cooling rate of about a million seconds per degree was sufficient to cool the five metre diameter mirror of the Mount Palomar telescope, which is made of a silicate glass.

Fundamentals

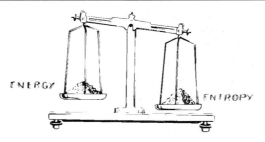

The Laws of Thermodynamics and Kinetics

Thermodynamics deal with the contrasting tendencies for atoms to seek either a well ordered, or a chaotic state of being. These states are usually discussed in terms of an internal energy, E, that tends to bring atoms together into close-packed (crystalline) formations, and a property called entropy, S, which tends to mix atoms into random formations, as in a liquid. The state that actually predominates is a function of the temperature and so the total energy, G, is given by:

$G = E - TS$

This is the equation for the combined first and second laws of thermodynamics. Matter always tends to be stable at its lowest total energy, as decided by the energy balance given in this equation. Thus at high temperatures, the entropy term dominates and the atomic structure is likely to be random; at lower temperatures the internal energy term dominates and the structure is likely to be crystalline.

Thermodynamics provides the driving force for a phase change to take place. The rate at which atomic shuffling can occur to bring about a phase change depends on the kinetics of the process, a highly temperature-dependent factor. Thus thermodynamics and kinetics have to cooperate closely if a phase transition is actually to occur.

Crystal transformations

An important characteristic of many crystalline materials is that they can undergo solid-state phase transformations from one crystalline phase to another. A phase transformation occurs during heating or cooling when the new phase, or new atomic arrangement, has a lower energy than the old one (see box). Thus the way atoms are packed together in iron and steel at high temperatures is different from the way they are packed at low temperatures (see Figure 11). The high-temperature, close-packed austenite phase of steel can also dissolve much more carbon than the low-temperature, less close-packed ferrite phase. The Japanese discovered this characteristic of iron centuries ago, and without really understanding why used this solid-state transformation to harden the edges of their swords by quenching the metal from red heat into cold water. Steel thus treated has a higher hardness because the carbon is retained in solution in the ferritic phase at ambient temperature, provided the quenching speed is sufficiently rapid to prevent the carbon atoms from diffusing and agglomerating.

Both metals and ceramics can undergo phase transformations indeed these transformations are very important to many of the properties of this crystalline group of materials. In contrast, poly-

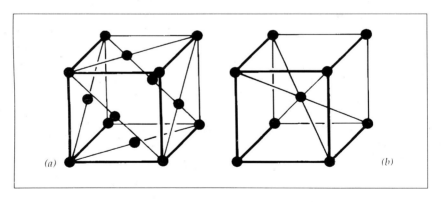

11 Two of the most common of the unit cells of metals: (a) face-centred cubic and (b) body-centred. Some metals, like iron, may be made up of both forms, in proportions that depend upon the temperature at which the iron is treated

Fundamentals

mers rarely undergo such solid-state phase transformations other than by the formation of thin, close-packed crystalline regions brought about by chain alignment or by chain folding. Other materials such as glass and glassy metals do not normally undergo phase transformations either, except by the process of crystallisation itself.

Composite and cellular materials

Not all materials are homogeneous solids. There are synthetic composites like reinforced concrete, and open cellular structures such as natural coral. It is worth examining these two types of material in more detail since they can differ substantially in properties from solid materials. They can also offer interesting advantages over solid materials in some applications.

Reinforced concrete and beyond

As illustrated in Figure 12, composite materials consist of at least two phases or components. This gives them properties different and in many cases superior to those of the individual phases. One of the most established of the composite materials is reinforced concrete, in which a matrix - an aggregate of cement and sand - is reinforced with steel wire or bar. The matrix is weak in tension but strong in compression, and it is also brittle, but the steel wire or bar is very strong in tension. The combination is a relatively cheap, tough material essential to the construction of large structures such as high-rise buildings, bridges, and oil platforms. A rather more sophisticated, yet nowadays commonplace, modern composite is glass-fibre-reinforced polyester, or epoxy resin. The glass fibres are extremely strong though relatively brittle, whereas the polymer matrix is weaker but relatively flexible, so that the combination of materials and properties results in a tough, strong material. This composite has many applications including small boat structures, skis, and automobile body construction.

Tomorrow's Materials

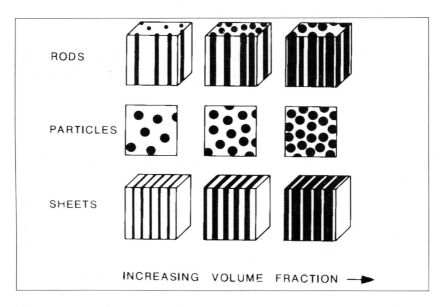

12 Composites consist of at least two components, or phases, in various proportions and shapes

A great deal of research is under way to develop different types of composites and material combinations. For example, fibre-reinforced concrete with improved casting techniques which result in much stronger concrete are now being studied. Another well-known composite family is the carbon-fibre-reinforced epoxy resins used in tennis racquets and aircraft structures (airframes). While these materials are highly anisotropic (stronger in one direction than the other), they are much lighter and stronger than conventional materials such as the best aluminium alloys.

Current composite research is concerned with producing cheaper and stronger fibres that are nevertheless compatible with their surrounding matrix. This means that the matrix material - which can be polymer, metal, or ceramic- must be compatible and have good bonding with the fibres it supports. Traditionally, thermoset matrix composites like epoxy or polyester have been used for advanced fibre composites since they are stronger than the thermoplastic matrix composites. However, the higher ductil-

Fundamentals

Table 1 Properties of various fibres used in composites

Property	Kevlar	E-glass	Carbon	SiC	Polyethylene
Mean fibre diameter, mm	2×10^{-4}	2×10^{-4}	10^{-4}	$1\cdot4 \times 10^{-4}$	5×10^{-6}
Density, $Mg\,m^{-3}$	$1\cdot44$	$2\cdot54$	$1\cdot8$	$3\cdot0$	$0\cdot97$
Young's modulus, $MN\,m^{-2}$	$1\cdot24 \times 10^5$	$7\cdot2 \times 10^4$	$2\cdot2 \times 10^5$	$4\cdot0 \times 10^4$	$1\cdot2 \times 10^5$
Tensile strength, $MN\,m^{-2}$	2760	3450	2070	4830	2590
Elongation to fracture, %	$2\cdot4$	$4\cdot8$	$1\cdot2$	...	$3\cdot8$

ity of the thermoplastic matrices make them more attractive for future applications. These 'advanced composite' thermoplastics can now compete with metals in terms of toughness, corrosion and wear resistance, and even heat resistance. A comparison between some of the fibres used in advanced composites is given in Table 1. While the tensile strength of carbon fibres is not outstanding, they possess exceptionally high stiffness, or Young's modulus as defined by the steep proportionality between stress and strain. High stiffness is a particularly useful property in materials used for airframes since it is essential that large deflections of the airframe are avoided.

A drawback with composites is that because of the sophisticated production techniques required they are expensive, and so the number of applications remains fairly modest. An example of the manufacture of glass fibre reinforced tubing is shown in Figure 13. The 'laying up' of carbon fibres in different directions for a composite aircraft skin is a more complicated procedure than that illustrated in Figure 13, and is usually carried out manually. Such a manual operation, however, would be too slow and hence too expensive for automobile production. Nevertheless, the advantages of using composites for lightweight structures are such that (provided fuel costs remain high) by the turn of the century composites are likely to command a substantial share of a market traditionally held by metals, particularly in the transport sector.

Tomorrow's Materials

13 A stage in the production of high-strength, wear-resistant glass-fibre-reinforced epoxy tubes. (Courtesy of Eivon Carlsson and ABB PLAST, Sweden)

A leaf out of Nature's book

Cellular materials are the most sophisticated of all materials, and are in many ways among the more exciting of tomorrow's structural materials. Cellular materials form a group separate from solid materials in that their properties, such as strength, stiffness, fracture characteristics, and heat conductivity, depend essentially upon their cellular structure. Thus the way these materials react to mechanical loading, for example, is determined by their cell wall dimensions and by the shape and density of individual cells, rather than just by the properties of the cell wall material

Fundamentals

itself. Most of these materials are also anisotropic, which means that cellular materials react to external loads in different ways depending on loading direction. An elegant example of a cellular material is natural cork (see Figure 14, p.26). The tiny cells making up the structure of cork are seen to be neatly stacked like house bricks. Cork cells were first observed by Robert Hooke through the compound microscope, his own invention, in the mid-1600s. Cork was one of the earliest naturally occurring materials to be used by man, and its application as flooring, roof insulation, fishing or swimming floats, and shoes, is mentioned in Ancient Greek and Roman literature.

Another familiar cellular material is wood. As shown in Figure 15, wood consists of an array of elongated cells, the walls of which are made of a composite of several layers of spirally wound cellulose fibres in a matrix of lignin. The arrangement of these fibres is such as to give the best combination of resistance to the many types of loading to which trees may be subjected. Indeed, wood is more intricate, by far, than any synthetic cellular composite currently available.

Attempts to imitate Nature's cellular structures remain on the whole fairly crude, but in spite of this there are several examples of cellular materials in everyday use. Examples are aluminium honeycomb structures that are used extensively in modern aircraft, helicopters and sporting goods, and paper honeycomb structures commonly used in packaging and for internal doors in houses.

Why metal bends, rubber stretches, and glass breaks

A well-known feature of plate glass is that, while it is fairly flexible when handled with care, the impact from a cricket ball or stone

Tomorrow's Materials

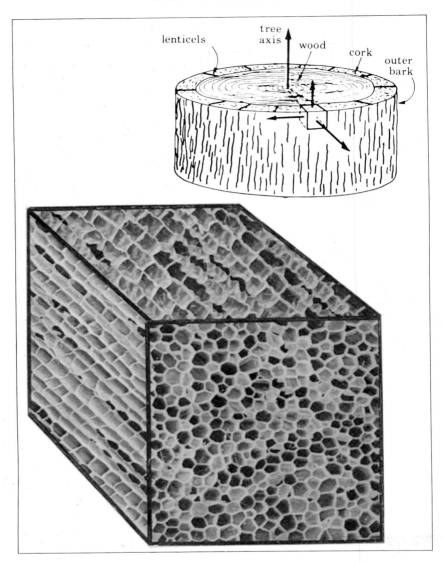

14 The microstructure of cork. There are approximately 250 000 000 of these cells in a single wine-bottle cork! The cellulose, suberin (a fatty substance), and waxes making up the cell walls of cork occupy about 10% of its volume; the rest is air. Its cellular microstructure gives cork many remarkable properties - for example, it is one of the best heat and sound insulators known. (Scanning electron micrograph by Ralph Harrysson, University of Luleå)

Fundamentals

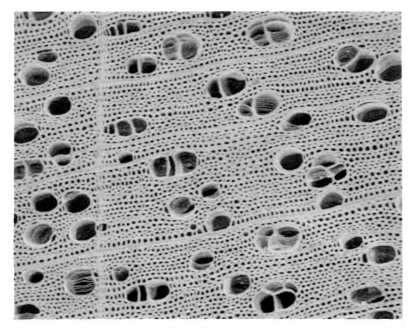

15 An electron micrograph of the microstructure of wood taken along the axis of the tree. The cells of wood are not as regular as cork, although the cell wall material (cellulose) is very similar. The elongated nature of wood cells in the direction of the tree axis makes this material highly anisotropic. Different types of tree have different cell wall thickness and different distributions of sap channels (the large cells) (Magnification x 75)

thrown at it is sufficient to shatter it easily. This is rather surprising in some respects since glass is inherently a very strong material and can, in principle, withstand high mechanical loading without breaking. Indeed, as it does not contain dislocations glass is virtually as strong as the bonds between its individual atoms. Crystalline metals are also inherently strong materials in that the bonds between atoms are very strong. However, metals are much less resistant to mechanical loading than glass since they contain dislocations which can be made to move relatively easily under stress. Amorphous metals do not contain dislocations in the conventional sense, and so their strength is also, in theory, as high as that of the atomic bonds themselves. However, there is some debate about this since amorphous metals have been found to contain 'slip bands' when deformed, and this implies that some sort of dislocation mechanism may be involved. An alternative view of amorphous metals is that they are so packed with dislocations that individual dislocation line defects are not readily identifiable. Further research into the characteristics of amorphous metals is needed to clarify this, however.

The mode of fracture in amorphous polymers depends on factors other than local atomic bonding strength. For example, if you break a plastic (PMMA) ruler in two, only about 10% of the fracturing is by molecular bonds (or chains) being broken. The rest is by individual chains being pulled bodily out of the material, rather like fibre pull-out known to occur in certain types of composites. In densely packed chain polymers of high molecular weight, chain pull-out occurs to a much lesser degree because the molecular chains are more tightly entangled. This aspect of polymer fracture partly explains the different speeds at which fracture occurs in these materials, in that chains require time to disentangle themselves if they are not to be broken up.

Some polymers, like rubber, are highly elastic, because of the way their polymer chains can adjust their position and shape with respect to one another. Likewise, crystalline phases in polymers are fairly rigid because there is little flexibility between the adjacent polymer chains, and the bonding of atoms across and along the chains is rigid and directional. Earlier types of elastic bands had the problem that, when stretched, polymer chains became aligned to some extent and a limited degree of crystallisation occurred, causing the rubber to become brittle (see Figure 16). This problem is overcome today by suitably modifying the polymer chain structure to prevent 'crystallisation' in the stretched elastic band.

The brittleness of glass is mainly due to the existence of surface defects or scratches. Glass breaks easily if a scratch is first made on its surface. The scratch acts as a stress raiser, and mechanical loading of the glass then causes the scratch to behave like a notch. When the glass is loaded the stress at the root of the scratch exceeds the bonding strength between atoms, and a crack is init-

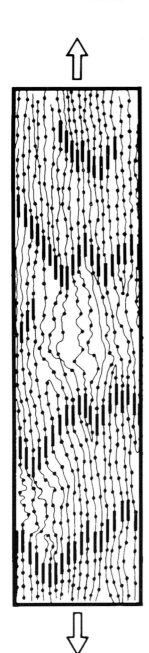
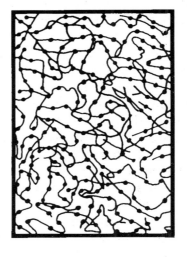

16 Stretching an elastic band can cause some alignment of polymer chains, resulting in 'crystallisation'. If this happens the elastic properties of the rubber deteriorate. The lower figure shows the polymer chains in 'relaxed' condition and the upper figure shows the stretched out chains

iated. Cracks find it much harder to grow in metals because stress concentrations, for example at the root of surface scratches, are effectively neutralised by dislocation movement or plastic flow.

Glass fibres, though, often exhibit an astonishingly high resistance to cracking when subjected to bending strains, and this is because they are relatively free of surface scratches. Plate glass can be cooled in such a way that compressive residual stresses remain at the surface, making it difficult for surface cracks to open up. Alternatively, large 'foreign' atoms can be forced into the surface layers of the glass by a process known as ion impregnation to achieve the required compressive stresses, and thereby hinder crack initiation.

So, the extremes of behaviour exhibited by different materials can be explained largely on the basis of their atomic structure. In a similar way, new materials can be 'designed' so as to have just the type of behaviour or property required.

Materials selection

It is a unique feature of metals that they can be used, melted down, and then used again - in other words recycled. This saves energy. About 40% of today's production of steel and 50% of aluminium is in recycled form, and this percentage will undoubtedly increase in future. However, there is the problem with recycling that what emerges the second time around is not necessarily the same product as the initial one. For example, scrap steel invariably contains many additional materials, such as copper from electrical wires in scrapped cars. But this has had some beneficial effects. For instance, heat-resistant nickel-based 'superalloys' were discovered by chance when materials containing 'unwanted' impurities like sulphur were found to resist high-temperature creep better than the uncontaminated version of the alloy.

There is no question that by the end of the century the amount of recycled material will have grown substantially. From an ecological point of view this is obviously a very good thing, but with high-grade metals the presence of unwanted impurities may

Fundamentals

cause difficulties. However, this problem is likely to be neutralised in future by improved techniques of analysis and better predictions of how materials behave in use.

Table 2 The energy necessary to produce a tonne of various types of material

Material	Energy required	
	GJ/tonne	Oil equivalent (tonnes)
Mild steel	58	1·5
Stainless steel	115	2·8
Brass	97	2·5
Aluminium	290	7·5
Magnesium	415	10·7
Titanium	560	14·5
Polyethylene	80	2·1
Nylon	180	4·7
PVC	80	2·1
Rubber (natural)	6	0·15
Rubber (synthetic)	140	3·6
Wood	1	0·03
Concrete	8	0·2
Floor tiles	9	0·23
Brick	6	0·16
Glass	24	0·6
Carbon-fibre composite	4000	103

Material properties

The properties of materials of most interest when considering structural applications are yield strength, stiffness (or Young's modulus), fracture toughness (or ability to resist crack growth) and density (mass per unit volume). In materials which will have to withstand high temperatures, the melting point is also of importance. The combination of high yield strength and good fracture toughness, or ductility, is what makes steel such an excellent structural material. When it is subjected to unexpectedly high loads (which happens more often than is generally appreciated), a steel structure does not break, but merely yields, redistributing its load over the structure as a whole. Plastics and ceramics

Tomorrow's Materials

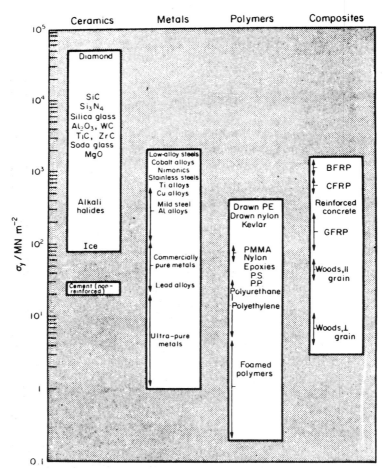

17 The strengths of different groups of materials. (From M. Ashby and D. Jones, 'Engineering Materials', Pergamon, 1980, p. 80)

cannot yield as they are quite brittle in comparison and this, together with the higher cost of these materials, means that structures made from them have to be designed much more carefully and accurately.

Figure 17 shows the yield strength of a number of different types of materials. Not surprisingly, ceramics in general have the highest strengths, with diamond the highest of all. The strength of steels and aluminium alloys is comparable to that of fairly weak

Fundamentals

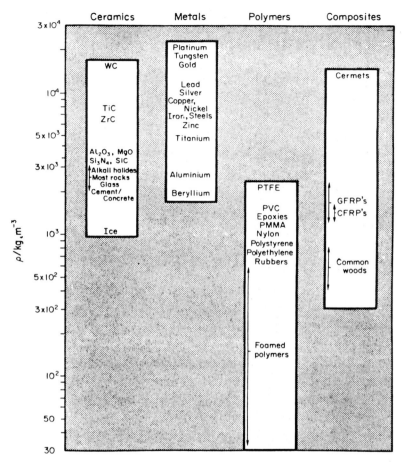

18 The densities of different groups of materials. (From M. Ashby and D. Jones 'Engineering Materials', Pergamon, 1980, p. 53)

ceramics. Carbon fibre-reinforced plastics and cement are about as strong as mild steel. Of the polymers, Kevlar fibres have strengths comparable to steel, while at the other extreme the foamed polymers used in packaging and furniture have very low strengths.

The densities of these classes of materials are compared in Figure 18. Ceramics are generally lighter than metals; polymers have the lowest density of all materials. Once again, composites

Tomorrow's Materials

Table 3   The role of design parameter on materials selection

| Material | Young's modulus, E (MN m^{-2}) | Density, ρ (Mg m^{-3}) | Specific Moduli* | | |
			$\dfrac{E}{\rho}$	$\dfrac{E^{\frac{1}{2}}}{\rho}$	$\dfrac{E^{\frac{1}{3}}}{\rho}$
Steel	210000	7·8	26923	59	7·3
Titanium	120000	4·5	26667	77	10·5
Aluminium	73000	2·8	26071	96	14·3
Magnesium	42000	1·7	24705	121	19·7
Glass	73000	2·4	30416	113	16·8
Brick	21000	3·0	7000	48	8·9
Concrete	15000	2·5	6000	49	9·5
Aluminium oxide	380000	4·0	95000	154	17·3
Silicon carbide	510000	3·2	160000	223	23·9
Silicon nitride	380000	3·2	120000	193	21·7
Boron nitride (cubic)	680000	3·0	227000	275	28·0
Boron fibres	400000	2·5	160000	253	28·2
Carbon fibres	410000	2·2	186000	291	32·3
Carbon-fibre composite	200000	2·0	100000	224	28·0
Wood	14000	0·5	28000	237	46·7

lie somewhere in between, for the simple reason that they consist of mixtures of different materials.

When comparing materials it is important not only to consider yield strength, stiffness, and density, but to express the property in terms of the material's specific modulus, or specific strength (see Table 3). The specific modulus is expressed as a power of the modulus, the power depending on the particular design condition or criterion used, divided by density. In this approach the designer finds how a component under stress will react by calculating its strength and stiffness as a function of its weight and the type of loading applied. Thus, for rods loaded in tension it can be shown that stiffness per unit density is the correct design criterion (E/ρ). On this basis there would be little to choose between steel, titanium, aluminium, or even wood for the structural part, as is apparent from Table 3, but there could be a significant advantage in using carbon-fibre composites instead. It is surprising that for

Fundamentals

panels under uniform pressure, where the design criterion is $E^{\frac{1}{3}}/\rho$ would apparently be the better material to use. This is because cellular materials like wood are both strong and light, and thus have a very favourable specific stiffness in such an application.

Obviously there are a number of factors to consider when designing structures, and the specific moduli are helpful in allowing engineering designers to compare different types of materials. Also to be considered is, of course, the cost of producing a material. For example, cost would rule out the use of carbon-fibre composites in place of steel and concrete for large structures, but in sophisticated products such as tennis racquets and critical components in airframes, carbon-fibre composites may have the edge over metals.

An interesting problem is how to reduce density (and thus weight), yet still achieve a very high strength and hence obtain the highest possible specific strength in structural materials. We have seen that in the natural world structural materials are usually in a cellular form; two well-known examples are wood and bone. So far we have not succeeded in producing anything like a cellular material to match wood or bone, but there have appeared a number of structural materials based on honeycombs made from aluminium. An example of the use of aluminium honeycombs for part of an aircraft wing is illustrated in Figure 19; most aircraft also have honeycomb materials for flooring and doors. Paper honeycombs are commonly used in doors in buildings, and in packaging. The very high specific moduli and strengths of cellular materials make it likely that these materials will find many more applications, particularly in the transport industry where there is a need to cut energy costs by reducing weight.

The extension to the available range of properties provided by natural and synthetic cellular materials is illustrated in Figure 20. It is seen that cellular materials complement the properties of solid and composite materials very well, in some cases (e.g. Young's modulus and density) substantially extending them. It seems likely that the future will see a much wider use of these interesting materials.

Tomorrow's Materials

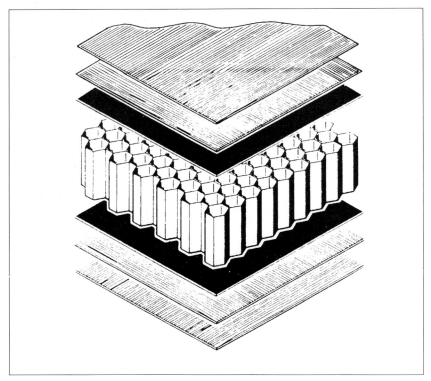

19 An example of the use of honeycomb materials in an engineering structure - part of an aircraft wing. These materials consist of a cellular structure of aluminium, polymer, or even paper, covered by one or more thin strong layers of a metal or fibre composite. The result is a lightweight material of very high stiffness, used in aircraft, skis, and door panels

Characterising Materials

By denting, pulling, squeezing, cracking, hitting with a hammer or scrutinising at them

Both designer and user likes to be sure that a material selected for a given role actually fulfils the properties specified by the manufacturer. The more demanding the role the more specific the characterisation needs to be. High-tech materials for fibre optics, semi-conductor chips or high temperature gas turbine blades are

Fundamentals

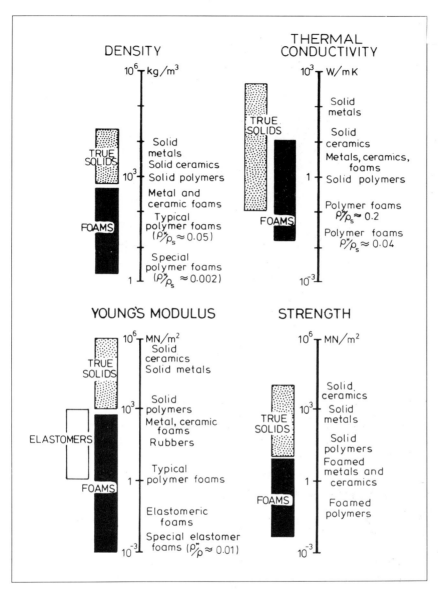

20 Cellular materials complement and considerably widen the range of properties of materials. From "Cellular Solids", by L. J. Gibson and M. F. Ashby (Pergamon, 1988).

Tomorrow's Materials

obvious examples where properties need to be up to specifications. But even materials used in bridges, oil-platforms, ships or submarines have to meet certain minimum requirements with regard to strength, toughness, weldability, corrosion resistance, etc. And if a part does fail in service, it is important to discover (and quickly) why, and what is to be done about it.

There are many ways to characterise and test a material; some of these are nondestructive and may, for example, involve measuring electrical resistance or surface properties by X-rays, or even resistance to corrosion in some difficult environment; others are somewhat more destructive and may involve denting the material with a diamond tool (hardness testing), pulling it out of shape (tensile testing), squeezing it like plasticine (compressive deformation), or boiling it in oil (heat treating). If these sound cruel, yet others may be positively destructive, like breaking the material open (crack-opening displacement), or smashing it with a sledge hammer (charpy impact testing).

Materials characterisation also involves scrutinising in a microscope to look for microstructural defects, measure changes in chemical species, isolate segregation, or identify inclusions. Nowadays microscopes do all this, and a great deal more. The type of electron microscope featured in Fig. 21, for example, can characterise practically anything from a computer chip to a turbine wheel and do it in at least a dozen different ways. For example, it can map out microstructures at magnifications up to a million times (with help from photographic processing); it can evaluate chemical compositions in several ways (by electron diffraction, microprobe analysis, wavelength spectroscopy, etc.); it can count and evaluate the amounts and types of different phases or inclusions using on-line computing; with the same computer it refines the quality of images, by subtracting unwanted electronic noise; it can store or record information in many ways (by laser disk storage, by still or video photography, etc.); it has such good depth of focus that complex fracture surfaces can be studied and mapped like atomic-level ordinance surveys.

Other types of microscope complement systems like the one illustrated, but work in quite different ways. A summary of a few

21 This scanning electron microscope is not unlike the control panel of a aeroplane. In fact, it can do everything except fly! The many keyboards and output consoles seen here help produce large amounts of data from any given specimen. Photo by Eivon Carlsson.

microscopy and microanalysis techniques (used to study advanced materials) is given in Table 4. Indeed, the analysis of microstructures and features or phenomena, ranging in size from atomic dimensions to metre sizes, are not only possible, but nowadays virtually routine.

In unusual applications, it is often necessary to combine advanced techniques of scrutinising with those of mechanical or electrical testing. As an elegant example of this, Fig. 22 shows a

Tomorrow's Materials

Table 4 Techniques for Microstructural Analysis of materials

Microstructural level (resolution range)	Type of microscope used	Complementary or on-line techniques for chemical analysis
mm to metres (10^{-3} m to 1 m)	light – optical[4]	X-ray diffraction and spectroscopy
micrometres to mm (10^{-6} m to 10^{-3} m)	scanning electron (SEM); scanning ion (SIMS)	electron microprobe (EDX[1] and WDX[2]) auger spectroscopy electron channelling
nanometres to micrometres (10^{-9} m to 10^{-6} m)	transmission electron[4] (TEM); scanning-transmission electron (STEM)	electron microprobe (EDX, EELS[3]) electron diffraction lattice imaging
single atoms (approx. 10^{-10} m)	field ion[5]; scanning tunnelling[6]	atom probe[5]

[1] Energy Dispersive X-ray Spectroscopy

[2] Wavelength Dispersive X-ray Spectroscopy

[3] Electron Energy Loss Spectroscopy

[4] The main difference between light optical and electron microscopes is resolution. This is due to the fact that the wavelength of electrons is much smaller than light (photons) and hence the resolution of electron microscopes is at least a hundred times better than optical microscopes.

[5] Field ion microscopes operate by pulling single atoms from a sharp pointed sample with the help of a strong electrical field. Each atom can be analysed for chemical species by passing it through a time-of-flight spectrometer.

[6] Scanning tunnelling microscopes are a relatively new toy of materials scientists, sampling atomic structure by measuring the tunnelling current between surface atoms and a sharp probe close to (but not quite touching) the surface.

For other definitions see the Glossary.

tiny micro-mechanical silicon device having one of its cantilever beams tested *within* an electron microscope. This device is so small it would not be possible to test it in any other way. The test showed that materials of this size behave quite differently from the bulk.

Fundamentals

22 Testing the elastic response of a micro-mechanical silicon device inside an electron microscope. The cantilever beam being tested has dimensions of 0.2mm wide x 0.02mm thick. Such beams are much "stronger" than the bulk material because they contain little or no defects. After work of Dr. Stefan Johansson, Uppsala University.

Further reading

Of the many excellent books on the fundamentals of materials science, five are particularly recommended.

R. J. Cotterill, 'The Cambridge guide to the material world' (Cambridge University Press, 1985). This book, containing many colourful illustrations, covers the entire range of organic and inorganic materials .

Tomorrow's Materials

J. E. Gordon, 'The new science of strong materials, or why you don't fall through the floor' (Penguin Books, 1976) and 'Structures, or why things don't fall down' (Penguin Books, 1978). These two inexpensive books make for enjoyable, often humorous reading and cover the entire range of materials science.

M. F. Ashby and D. J. Jones, 'Engineering materials' (Pergamon, 1980). This book is for the more technically inclined and gives a good coverage of the interaction between materials science and engineering design.

A. H. Cottrell, 'Mechanical properties of matter' (John Wiley, 1963). This outstanding book is regrettably out of print, but it can be obtained at any good technical library. It is arguably the best book ever written on the subject of materials science, but is definitely for the scientific minded only.

PART II
APPLICATIONS

This second part takes a look at some of the many exciting new materials now under development. These include new lightweight aluminium alloys and fibre-polymer composites for aircraft skins, rolled structural beams made of toughened concrete, new polymers that may soon displace metals, advanced ceramics that will revolutionise the machine-tool, electrical and automobile engine industries, the most transparent windows in the world (fibre optics), new generations of transistors, revolutionary superconducting oxides which promise to change our way of life, and high-tech materials that upgrade our sporting performance.

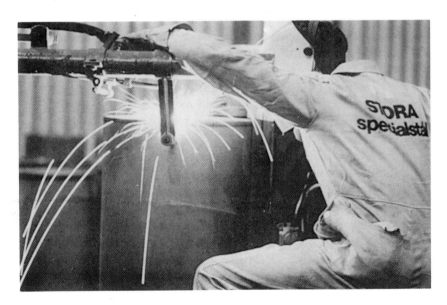

Selected wonders of the world of advanced materials 1
High strength low alloy (HSLA) steels

These materials are remarkable for the fact that they are sophisticated and yet are produced in millions of tonnes per year. They are in fact the most important material used in the construction of large structures such as bridges, ships, oil platforms and cranes. The high strength and good weldability of HSLA steels is due to their very fine grain size, typically around ten micrometres. This is achieved by very small, highly controlled alloying additions to the iron of elements that combine during hot rolling of the cast metals to produce an extremely fine distribution of tiny precipitates. These precipitates are rather complex but very stable intermetallic compounds usually based on at least two metallic components such as niobium, aluminium, titanium, or vanadium, and two non-metallic components like carbon and nitrogen. These 'low alloy additions' typically comprise only 0.15% of the total weight of the steel. The mean size of the particles is about 50 atoms in diameter and they can only be seen under a powerful electron microscope. In spite of their small size, however, these particles are very efficient in 'pinning' grain boundaries, thereby hindering grain coarsening from occurring during thermal treatments like normalising anneals or welding. The micrograph illustrates one of these tiny particles holding up the movement of a grain boundary in steel, like a fish caught in a drifting net, as observed through a transmission electron microscope.

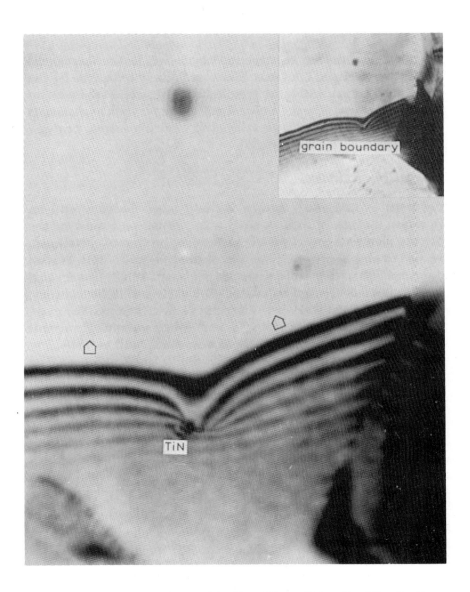

Transmission electron micrograph by Simon Ringer, University of New South Wales

Tomorrow's Materials

Structural materials

The two most important structural materials in the developed nations are steel and concrete. This is still likely to be true by the turn of the century, although plastic structures are on the way up and wood may well make a comeback. Here we look at materials used in large structures, such as ships, oil platforms, bridges, roads, railways and buildings.

Steel getting stronger

It is hard to imagine a future without steel. There is still an abundant supply of iron in the Earth's crust, and most of the alloying elements used in steel, with the exception of chromium, are also in ample supply. The current annual production of steel is around 700 to 800 million tonnes per year. While this level is likely to have decreased by some 20% by the end of the century, more sophisticated steels will continue to be developed for making structures of higher strength and lower weight. If present trends are anything to go by, these new steels will be low-alloy fine-grained materials with good weldability and toughness. The medium-strength 'clean' weldable steels for pipelines will almost certainly continue to be manufactured throughout the coming decades, on about the same scale as today, but it seems likely that aluminium alloys and carbon-fibre composites will make significant inroads on steel in areas such as automobile body production. In addition, the high-alloy steels currently used in applications where a good wear resistance is necessary are likely to be progressively replaced by ceramic-coated materials. Less stainless steel will be produced as chromium becomes more scarce, and steels of the type used for example in the chemical industry may well be replaced by titanium or coated ferritic steels. Nevertheless, the excellent all-round properties of steel will ensure that this material continues to be used for many large-scale structures such as ships, oil platforms, pipelines and buildings, and as reinforcements in concrete.

Applications

Modern high-strength low-alloy steels have a fine grain structure, which provides both high strength and good crack growth resistance, or fracture toughness. These properties are achieved by selecting a manufacturing method which gives the steel a uniform distribution of very small and chemically stable particles (carbides or nitrides of approximately stoichiometric composition) which effectively 'pin' or hold the grain boundaries in position. The manufacture of these very sophisticated steels has changed considerably over the past decade. In particular, the process known as controlled rolling has attracted much interest. Steel is rolled to plate at a temperature high enough for the steel to be relatively soft, but at the same time sufficiently low to cause fine carbonitride particles to precipitate out. On cooling the steel, these precipitate particles hinder excessive growth of austenite grains, so that at the end of the cooling period, following transformation to the low-temperature ferrite phase, the plate has the required fine grain size. Since yield strength is proportional to the square root of the grain size, fine-grained steels are stronger than coarse-grained steels. This hot rolling process is illustrated in Figure 23, and an example of the fine dispersion of carbonitrides present in a structural steel is shown in Figure 24.

Another way of making high-strength steel is to quench it rapidly from its high-temperature austenitic phase to ambient temperature, which produces a hard martensitic microstructure (a solid solution of carbon in iron). The steel is then tempered to produce a high-strength alloy of good toughness. This tempering heat treatment causes fine particles of carbide to precipitate in the steel, pinning dislocations and grain boundaries. For applications in which a very high ductility is needed, 'clean' steels have been developed in which there are very few interstitial alloying atoms, such as carbon and nitrogen, to strengthen the material. These steels are therefore suitable for deep drawing or severe forming operations, and in their deformed state possess very high strengths because of the high densities of dislocations produced.

Steels can be treated in this way because iron has the unique characteristic that a phase transition, from one crystal structure to another (austenite to ferrite), occurs on cooling. With a knowledge

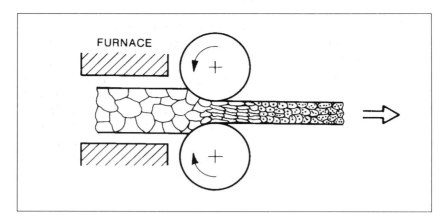

23 Controlled rolling, one of the modern techniques for producing fine-grained high-strength steels. After hot rolling the deformed structure recrystallises into fine grains which are prevented from growing by the simultaneous precipitation of extremely small carbides and nitrides. Cooling to ambient temperatures gives a very fine-grained ferritic structure with a strength of up to 600 MN m-2

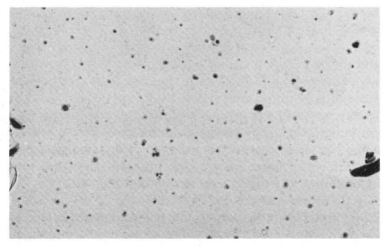

24 An electron micrograph showing fine carbides and nitrides within a single grain of a ferritic steel. (Micrograph by Jan Strid, University of Luleå

(Magnification x 100000)

Applications

of this phase transition and the characteristics of carbide precipitation, it is possible to manipulate the microstructure of steels to produce a wide range of properties.

A completely new type of steel, again based on taking advantage of this phase transformation, has recently been developed in Japan. This steel contains a dispersion of micron-sized oxide particles. As the steel cools after hot rolling, or welding, these oxides act as nucleants for the ferrite phase. Thus if the high temperature (austenite) phase is very coarse-grained, the final (ferritic) microstructure is very fine-grained and strong.

Welding metals

All large structures built from steel beams must be welded. This is a process in which an intensely hot electric arc is used to melt the beams locally and fuse them together usually with the aid of a 'filler' metal of similar composition to the beam material. In fusion welding the temperature changes over a very wide range in a very short time, bringing about marked changes to the microstructure and properties of the base material. It is unlikely that fusion welding as a means of joining large-scale structures will be replaced in the coming decades by other techniques such as gluing or diffusion bonding. Enough is understood of the physics and physical metallurgy of fusion welding for weld cracking to be avoided. For example, welding engineers can now use a micro-computer to predict the size and shape of the heat-affected zone for a given welding process and material composition (see Figure 25). The increase in grain size near the fusion line in Figure 25 is the main cause of property changes since cracks grow more easily in coarse-grained material. These microstructural changes may weaken the welded joint further if hydrogen, picked up in the weld from moisture or dirty surfaces, enters the steel.

Stainless steels

Corrosion is one of the most serious problems with metals such as steel. One solution has been the development of a range of so-called 'stainless' steels, each tailored to a specific application: fully austenitic steels for kitchenware and chemical plant, ferritic stain-

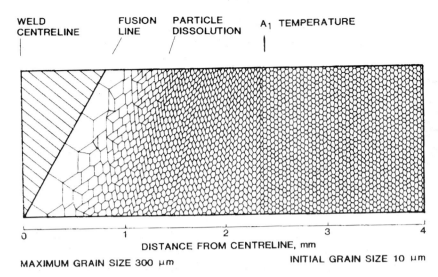

25 A computer-generated image of the heat-affected zone in a welded steel. The intense heat from the electric arc has modified the original fine-grained structure, producing considerable grain growth. The coarse-grained zone of the weld is potentially troublesome in that it reduces the toughness of the joint. The fusion line represents the boundary between solidified and base metal, and the 'A1 temperature' label marks the place in the weld where the ferrite-to-austenite phase transformation occurred when the plate cooled after welding. (Computed image generated by Fred Scott, University of New South Wales)

less steels for automobiles, martensitic stainless steels for hard-wearing machine parts, and so on. However, these steels have certain drawbacks in that they are rather expensive because of their high alloy content. The main alloying elements used are the fairly scarce metals chromium and nickel, which together make up some 25% of the steel. At temperatures above about 900°C the corrosion resistance of these steels fails, which can be a problem in certain types of chemical plant and in applications such as heating coils in furnaces. At present, this is overcome by surface treatments such as electrodeposition, and direct alloying or cladding with the help of heat treatments by laser beams. As mentioned above, it would be tempting to replace stainless steel with titanium alloys in chemical plant because of their superior corrosion resistance, but their cost, compared to stainless steel, is prohibitively high. Use of this valuable light metal is currently

Applications

confined mainly to aerospace and sports applications, but this situation may change in the future as the metal finds more applications and becomes cheaper to produce.

Tougher concrete

Concrete is one of the most widely used structural materials, being relatively cheap and very good at bearing loads. Recent research and development has concentrated on improving the tensile strength and toughness of concrete. The results of these improvements are soon likely to be seen in concrete structures such as oil platforms, which are subject to cyclic loading.

Up to now the poor mechanical performance of unreinforced concrete has precluded its use in applications, currently dominated by metals, where good tensile strength is required, but this may well change in the near future. Recent studies have shown that the low tensile strength of cement paste results mainly from the presence of microscopic pores. A typical microstructure of reinforced concrete, illustrating the presence of pores and other defects, is shown in Figure 26. The microstructural feature of concrete that distinguishes it from ceramics is the wide range of pore sizes. It is now known that the major factor determining the strength of cement is the size of these larger pores, which result from poor particle packing and air being trapped during solidification. Indeed, if porosity is eliminated from cement then high flexural strengths, comparable to that of mild steel, can be achieved.

Various ways have been tried to reduce the porosity and improve the strength of cement. One obvious approach is to use mechanical treatments such as vibrational compaction. Cement-dispersing compounds, which effectively reduce the amount of water that needs to be added so as to give the cement paste a workable consistency, have also been used. The filling of pores with solid materials such as sulphur and resin, and the use of fine-grained cement and fibre-reinforcing materials are yet other approaches.

Perhaps the most promising development for improving the tensile strength of concrete is one in which water-soluble organic polymers are added to the cement/water mixture. The polymer/

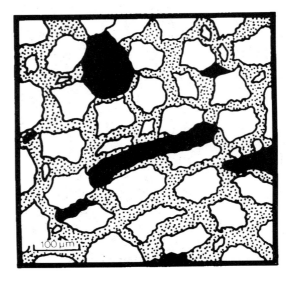

26 A schematic illustration of a typical microstructure of concrete, with considerable porosity (black areas). It is this porosity that gives concrete its poor tensile strength

water composition alone gives a fairly stiff 'gel' which effectively forms a deformable 'dough'. Dispersed in this dough, and occupying about 60% of the volume, are cement particles, and the polymer/water gel effectively fills any excess space left between the cement particles. The resulting material is then processed by conventional plastic deformation/press-moulding techniques, such as extrusion and rolling. Once the material is formed into the desired shape, it hardens as normal inorganic hydration reactions take place within it. The final densified material thus consists of polymer-bonded, close-packed cement grains containing little or no residual porosity as a result of the filler and the type of forming processes used. The resulting microstructure is illustrated schematically in Figure 27. Other filler materials, such as glass fibres, silicon carbide and alumina particles or fibres, have been used in the polymer/cement mixture. In this way materials with a wide range of properties can be obtained. An example of a "tough" tensile fracture of cement reinforced with tiny bundles of glass fibres is illustrated in Figure 28.

Applications

27 A schematic illustration of the microstructure of a newly developed concrete in which the spaces between the hard grains contain polymer filler. The bending strength of this type of concrete compares well with that of mild steel

It seems likely that the time is not far off when relatively high-strength beams of concrete in the form of pipes, rods, T-sections, thin sheets and even complex shapes will be in production. Complex extrusions and press-mouldings of concrete will soon be delivered from the 'concrete mill', to be assembled on site just as steel is today. Under development are new polymer-based glues capable of bonding together the component concrete parts, and similar to the types of polymer already used to bond the individual cement particles.

Wooden elegance

Wood is in many ways a very elegant structural material. It is cellular, with a high specific modulus which makes it more attractive for certain applications than synthetic materials (see Part I; material properties). Wood is still the most widely used of all structural materials in the world. This is mainly because in Third World countries it is cheap and accessible, but these attrib-

Tomorrow's Materials

28 Glass fibre reinforced concrete. This scanning electron micrograph shows a "tough" fracture in the tensile surface of a bend specimen. After work of Dr. Les Henshall, University of Exeter.

utes have led to overuse and have caused it to become quite scarce. Consequently a number of centres for wood research have been established, and are now rethinking the future role of wood in structural engineering. As with other expensive and sophisticated materials, it is now apparent that more thought must be given to how wood can be more effectively utilised in the future.

Materials scientists now regard wood as an advanced composite material. Its aesthetic beauty and excellent structural properties make it ideal for structural beams in buildings (see Figure 29).

Applications

29 This photograph of composite wooden support beams illustrates the aesthetic beauty and structural importance of wood

The beams in this photograph are built up from separate units in order to increase fracture strength as well as develop their overall form. Wooden units are easy to glue together - yet another advantage of this material. It is even claimed by architects that fire is no more a hazard with wood than with steel beams: as wood is cellular it has poor thermal conductivity, so even when its surface burns, its bulk retains structural strength; metal, on the other hand, being a good conductor, heats rapidly and softens.

Other wood composites under continuing development are plywoods. There are many different varieties of plywood nowadays, providing a growing range of properties and applications. Methods of hardening and preserving wood are also receiving much attention, giving yet more breadth of application to this remarkable material.

Lightweight materials

To move anything or anybody requires energy. Whatever the mode of transport, materials used in the construction of vehicles need to combine good structural strength with high resistance to

Tomorrow's Materials

Table 5   Applications of aluminium. (After P. F. Chapman and F. Roberts, 'Metal resources and energy', 1983, Butterworth)

Use	Percentage
Beer and soft drink cans	40
Electrical	12
Wire	5
Motor vehicles	24
Aircraft	3
Saucepans	3
Building, roofing	8
Packaging	3
Metals Industry (alloying, powders)	2

corrosion, and both they and any containers they carry must be as lightweight as possible in order to assist handling and save fuel. In principal these requirements apply equally well to cars, aircraft, railway carriages, and food and drink containers. It would seem, then, far better to use light materials such as aluminium and titanium alloys, plastics, and composites rather than the heavier steels and concrete. Unfortunately, as we have seen, most light materials like aluminium, titanium, zinc, beryllium, magnesium, and the engineering polymers and composites cost more to produce than steel, wood and concrete. This means that special requirements of cost-effective design have to be employed if these materials are to rival steel or concrete. The term 'cost-effective design' is being used more and more in conjunction with advanced high-cost materials. The implication is that components made of such materials have to be designed to use the least possible material in the most effective way. This requires far more precision in both design and manufacture than is necessary when using conventional materials.

Aluminium and the light metals

Aluminium is nowadays used for an enormous range of products (see Table 5), including buildings and roofing, kitchen utensils, and soft drink and beer cans. More sophisticated aluminium alloys are used for skins and frames of aircraft. Cast alloys of

Applications

aluminium and silicon are widely used for automobile engine block materials, as they are strong, light, and have good corrosion and wear resistance. Titanium is more expensive to produce than aluminium, but has a higher specific modulus and is more heat resistant; in addition it is more corrosion resistant than stainless steel. Although the metal is expensive to produce, titanium and its alloys are used in a growing number of applications today including gas turbines and highly loaded components in airframes, and also in the chemical industry as a replacement for stainless steel. Zinc and magnesium are used as lightweight casting materials and for hardening high-strength aluminium alloys. Recent uses of magnesium castings include bicycle frames and gear box castings for helicopters. Indeed, magnesium casts so well that it is likely to be used more in future, particularly if its corrosion resistance can be improved.

Beer cans

Beer and soft drink cans provide an enormous market for aluminium, currently swallowing at least 40% of this metal's worldwide production. Although only a relatively cheap (low-alloy) grade of aluminium is used, the production of these cans makes use of some fairly advanced materials science. Each can is stamped out from a small 5 cm diameter disc, by a deep drawing process designed to utilise the maximum possible amount of ductility that can be squeezed out of the metal. The deep drawing operation gives the elongated grain structure a certain 'texture', or common crystal orientation, rather like an aligned fibre composite. The fibrous structure, together with the very high dislocation density of the deformed material, provide the can with adequate strength against buckling when handled, in spite of its paper-thin walls. It can be shown that each cubic millimetre of can contains some hundred million dislocations, and this alone gives it considerable hardness. It is an interesting fact that about 50% of every can has experienced a previous existence as can material, as a result of aluminium recycling. Indeed, there appears to be every likelihood that this particular form of metal reincarnation may continue well into the next century!

Tomorrow's Materials

Aircraft and spacecraft

Aluminium alloys have been used as airframe materials for around 75 years. By what is said to have been an accidental discovery, it was found that an aluminium alloy containing a little copper and magnesium, while not quite as light as pure aluminium, was considerably stronger. The alloy was called Duralumin, and several production aircraft based on this material were produced by Junkers during World War I. In spite of the fact that airframe materials have been the subject of an enormous amount of research over subsequent years, present-day aircraft are still built from a very similar type of alloy. Duralumin is actually a very sophisticated alloy in which extremely fine precipitates of particles rich in copper and magnesium, one to two atoms in thickness, harden the aluminium matrix. Recently it has been found that the properties of aluminium alloys can be improved considerably by including small additions of lithium. Lithium is the lightest of all metals, so that by using 2-3% by weight of this element the final alloy is slightly lighter than Duralumin, and it also has a higher stiffness. Although 2-3% does not sound very much, a large modern aircraft like a Boeing 767 consists of up to 80% by weight of aluminium, so tens of thousands of litres of fuel per year would be saved by replacing the present alloy by an aluminium-lithium alloy. The future is likely to see even larger lithium contents in aircraft alloys, up to 5% by weight, which will bring further improvements in strength, weight reduction, and fuel saving. The change from conventional Al-alloys to Al-Li alloys has already been made on Westland's EH101 helicopter (see Figure 30), and is planned for the Airbus.

Aircraft are among the few large scale structures that are not fusion welded. Fusion welding would modify the microstructure of these sophisticated precipitation-hardened metals to such an extent that the resulting change in properties would bring about an increased risk of stress-corrosion cracking and a lower fatigue resistance. However, welding processes such as laser and electron beam welding can now produce welds with extremely small heat-affected zones, and current research is showing that welds can be

Applications

made in such a way that the heat-affected metal precipitates and hence hardens as the weld cools.

In modern aircraft the proportion of aluminium alloys used is actually decreasing: they are losing out to composites and other

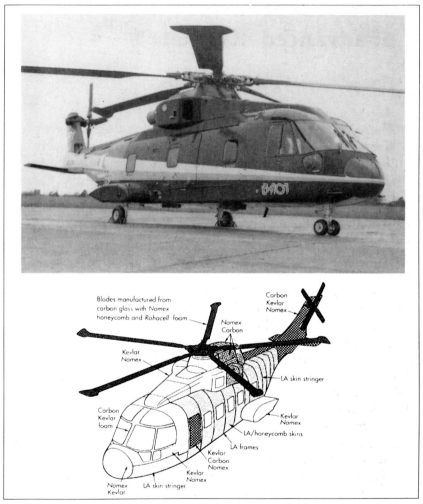

30 Westland's EH101, shown here, is the first aircraft to use exclusively aluminium-lithium for its main frame and skin. Note, however, the impressive range of polymer-based composites employed. "LA" in the figure refers to 'light alloy'. Courtesy of Westland Helicopters Ltd.

Selected wonders of the world of advanced materials 2

Fibre reinforced polymer composites

These materials are remarkable because they are very strong, possess high stiffness, and yet are extremely light. As a composite, they combine the enormous uniaxial strength of, for example, fine carbon fibres and the lightweight but brittle matrix of polyester or epoxy, or other polymer matrix materials binding the fibres together. They are used as structural materials in aeroplanes and spacecraft, and in certain sporting equipment such as skis and vaulting poles. In a way, advanced composites in cellular form are synthetic copies of structures found in Nature, such as the leaves of the lily plant or the wings of the dragonfly. Lily leaves possess good mechanical strength and lightness because they are constructed in the form of tiny cells bordered top and bottom by fibrous cellulose skins. The wing sections of aircraft are similar in construction, consisting of cells of a fibre reinforced polymer or thin aluminium honeycombs, bordered by thin fibre reinforced polymer skins. Such a composite is very rigid and light in weight, and yet possesses good toughness and very high strength. Ideal characteristics in other words for use in aircraft or spacecraft. The figure is an electron micrograph of the cross-section of a fibre reinforced composite of a type used in aeroplane construction.

Applications

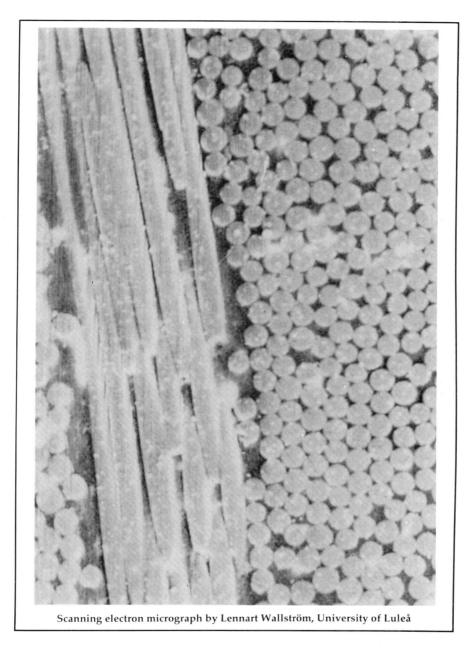

Scanning electron micrograph by Lennart Wallström, University of Luleå

Tomorrow's Materials

materials (see Figure 31). The distribution of composites in a modern jet fighter is illustrated in Figure 32. Military aircraft normally set the trend for developments in aviation, and civilian aircraft will probably consist of at least 30-40% by weight of carbon-fibre composites by the turn of the century. There are already experimental aircraft in use in which the skin is made entirely of carbon-fibre composites, giving substantial weight reductions over conventional metal aircraft. Honeycomb materials, consisting of aluminium cores and carbon-fibre composite skins, are also being increasingly used in aircraft structures. The Voyager aircraft which recently circumnavigated the Earth non-stop represents the ultimate application of these materials in aviation - its entire structure is made from cellular polymer composites, producing a superlight craft of high stiffness and strength.

Other composite materials being investigated for use in aerospace structures include aluminium-containing silicon carbide ceramic fibres and carbon-fibre-reinforced PEEK (a very tough polymer matrix material based on polyether etherketone). These toughened polymer-matrix composites are expensive, but are less prone to cracking than the epoxy and polyester materials currently in use. Such developments are typical of modern materials selection, in which traditional materials like the aluminium alloys have to compete with sophisticated new composites and honeycombs.

Engineering polymers

Polymers form the basis of many natural lightweight materials, such as wood cellulose, starch, resins, and proteins. By studying these natural polymers materials scientists have not only been able to gain a good understanding of their molecular make-up, but in many cases have succeeded in producing synthetic versions of them. Today, polymer-based materials command a large share of the total materials market. However, many polymer scientists feel that the market for conventional plastics is just about saturated, and that further expansion can be achieved only

Applications

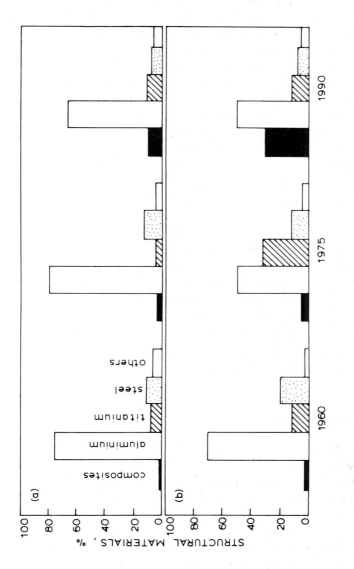

31 Materials used in (a) civilian and (b) military aircraft, 1960-1990. (Courtesy of C. J. Peel, Materials Science and Technology, 1986, volume 2, pp. 1169-75)

Tomorrow's Materials

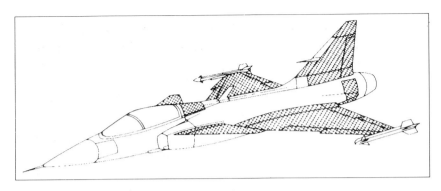

32 Fibre composites (shown hatched) are used extensively in this modern military aircraft, a Saab JAS. Civil aircraft are rapidly following suit

by developing more sophisticated products capable of pushing into areas traditionally occupied by metals. One of these areas, as we have seen, is composite materials, for which new polymers are being developed for use as both fibre and matrix. Indeed, polymer-based materials make up the great majority of today's composites. But composites are still only a small part of a much broader class of advanced polymer products generally known as engineering polymers.

Manufacturing plastics

To produce a plastic, a whole bundle of polymers - rather like a pile of cooked spaghetti - is mixed with various fillings, plasticisers, stabilisers, and pigments. This mixture is shaped as desired at a temperature high enough to keep it fairly soft and then cooled into its final rigid form. Table 6 gives the more important polymer types and the plastics that are made from them. Plastics are not all polymer: polyvinyl chloride, for example, used in floor tiles, contains only about 50% polymer, the rest consisting of a filler material.

Of all the plastics used today, 95% are based on as few as four main polymer types, although there are hundreds of variations of each main type. These four polymers are:

Applications

(1) polyethylene (polythene), in both solid and foamed forms, used for example in packaging, kitchen bowls and pails.
(2) polypropylene, used for example in automobile fittings
(3) polystyrene, used for example in household appliances
(4) polyvinyl chloride (PVC), commonly used for example as plastic piping in kitchens and drains.

Of the four, polyethene is the most common, with a current annual production of over 25 million tonnes.

Polymers are very versatile materials and their component chains can be knitted to produce many shapes and forms, as shown in Table 6. They can be packed together at random or interspersed with crystalline regions, depending on the method of production. These regions in turn contain linkages, called cross-links, that extend like bridges between domains. The cross-links provide extra stiffness, heat resistance and toughness. It is because of the various combinations and linkages that can be manipulated by polymer scientists that so many different types of plastics are available, and also why new forms of plastics continue to pour on to the market. Examples of new polymer types that have appeared in recent years include the tough polyether etherketone (PEEK), and the liquid crystal display (LCD) polymers. These contain very stiff chains regularly arranged, as in crystalline polymers. This 'crystalline' form is currently used as a display matrix in compact TVs and microcomputers.

Radically new polymer types based on totally new molecular forms are unlikely to appear in the coming decade or so. Apart from the scientific problems of producing them, it is now extremely expensive to launch new, unknown products into what is a fairly conservative materials market. There is probably more money to be made in developing speciality polymers which, by having better properties than some established materials, might push their way into new areas of application. One promising development along these lines has been to blend, or 'alloy', two or more different types of polymer. Today some 20% of all polymer production is accounted for by polymer blends, and injection-moulded polymer blends is one of the fastest-growing areas in the plastics field.

Table 6 Polymers: structural characteristics and applications. (After H. F. Marks, special issue 'Materials', *Scientific American*, 1967, and courtesy *Scientific American*)

Polymer characteristics	Structure	Examples	Uses
Flexible and crystallisable chains		Polyethylene	Pails, pipes, thin films
		Polypropylene	Steering wheels
		Polyvinyl chloride	Plastic pipes and sidings
		Nylon	Clothing
Cross-linked amorphous networks of flexible chains		Phenol formaldehyde	Television casings and telephone receivers
		Cured rubber	Tyres, transport belts, and hoses
		Styrenated polyester	Finish on automobiles and appliances
Rigid chains		Polyamides	High-temperature insulation
		Ladder molecules	Heat shields
Crystalline domains in a viscous network		Terylene (Dacron)	Fibres and films
		Cellulose acetate	Fibres and films
Chains moderately cross-linked, with some crystallinity		Neoprene	Oil-resistant rubber goods
		Polyisoprene	Particularly resilient rubber goods
Rigid chains, partly cross-linked		Heat-resistant materials	Jet and rocket engines, and plasma technology
Crystalline domains with rigid chains between them and cross-linking between chains		Materials with high strength and high temperature resistance	Buildings and vehicles

Applications

Polymer alloys are not to be confused with co-polymers (sequential chains of two or more types of molecule), which have been in existence for some time. They are also quite different from composite materials, in which the different materials or phases are distinguishable on a relatively coarse scale. The special characteristic of polymer alloys is that a polymer of one type is blended with that of another on such a fine scale as to be almost a single homogeneous phase - not unlike the single phase of a metal. To produce a polymer alloy an expensive polymer of one type with, say, good strength but poor formability, may be mixed with another inexpensive but brittle type which has good formability, with the aim of achieving a combination of good strength and formability. However, for this approach to work much depends upon the compatibility of the mixture as a whole. Like mixtures of water and oil, many polymers are simply not compatible with one another. As one observer has pointed out, getting two polymer molecules to alloy is like shaking two balls of wool together and hoping they will knit a scarf! The ingenuity of the polymer scientist is pushed to the limit in trying to devise blends that are both compatible and economical to produce. An alloy of this type, first produced successfully some years ago, is denethyl phenylene which goes under the tradename Noryl. It is made by alloying a polymer (PPO) which has good strength, but poor formability unless heated above 200°C, with polystyrene, which has poor toughness but good formability. The resulting blend, which can be formed at temperatures as low as 140°C, has good impact toughness and formability and is commercially successful.

There are currently dozens of new polymer alloys being investigated by polymer scientists. Among the more promising is an alloy consisting of blocks of the crystalline diphenyl siloxane and the rubbery dimethyl siloxane. This so-called block co-polymer, shown schematically in Figure 33, behaves like a thermoplastic elastomer. An attractive feature of this particular alloy is that it retains its rubbery properties over an impressively wide range of temperatures (between -50°C and +100°C). Furthermore, by varying the relative proportions of the two polymers, and with suit-

Tomorrow's Materials

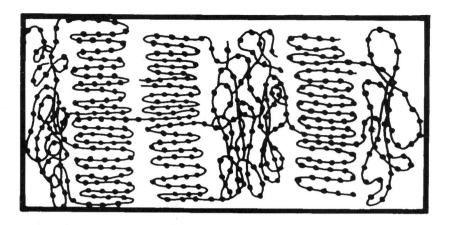

33 Higher strengths are achieved in block co-polymers in which hard crystalline particles are dispersed in a rubbery matrix

able heat treatment, the alloy can be made to behave as a tough elastomer or a high-stiffness material. Block copolymers are currently being evaluated as matrix materials in carbon-fibre composites where good impact resistance is needed.

An exciting development in block co-polymers has been the production of multiphase alloys in which amorphous and crystalline phases co-exist as domains, or ultra-fine particles, in an amorphous matrix. Since the atomic bonding in the domains is ionic, the blends are referred to as 'ionomers'. These alloys are extremely tough materials in that they can withstand high-impact loading without breaking up. They are also creep resistant, and so retain their shape even when exposed for long periods to moderately high temperatures (around 100°C). Commercial examples of this type are 'Surlyn' and the ionomeric polythylene. Another example of an ionomer alloy is a polystyrene-based blend containing sodium methacrylate as the ionic constituent. With further improvements in toughness and creep properties, these light and corrosion-resistant ionomers could well begin to replace some metallic structural components within the coming decade or so.

Applications

Plastic cars and bicycles

Engineering polymers and polymer-based composites account for 20% (by volume) of the material in a modern car. While it is the more sophisticated polymers based on carbon-fibre composites that are becoming more widely used in aircraft, cheaper forms of engineering polymers are being introduced in mass-production industries like car and truck manufacture. These engineering polymers are relatively easy to produce in finished form, and are also lightweight and resistant to corrosion, but raw material costs are substantially higher than those of steel. The cost of the steel strip that leaves the rolling mill is only a tenth of its value in the finished product, but with polymers the material costs about a third of its final value in the car. So if polymers are to replace steel in the automobile industry, not only must they have satisfactory properties but also, taking all operations into account, they must be cheaper than steel (unless they have other benefits). To achieve this, a new approach to the design and production of polymer parts is necessary. For example, the possibility of producing large, aerodynamically shaped, injection-moulded polymer car bodies (e.g. from polyurethane) in just one forming operation would provide serious competition to steel car bodies, which require several stampings, spot welding, and corrosion protection treatment to achieve a comparable final shape and properties. And polymers are not being used only for car body materials: other applications include springs made from an elastomer-based composite, carriage (leaf) springs of fibre glass, nylon fan wheels, wheel arches in polypropylene, and piston rings and skirts made from the new tough thermoplastic PEEK. In addition, polymer-based composites are being tested as axle materials, while moulded microcellular polymers (polyurethane foam) are already in use as bumper material.

The first all-plastic bicycle, recently produced in Sweden, is shown in Figure 34. Here again, new design and manufacturing concepts were used and the finished article, while not looking exactly like a traditional bicycle, at least gives steel bikes a run for their money! Its wheels are of a glass-fibre-reinforced polyamide

Tomorrow's Materials

34 The first all-polymer bicycle. The unconventional design exploits the good formability of polymer composites. See also p.129

and the frame is of a glass fibre/polyester, both of which are injection moulded. Even though this composite costs much more than steel, the use of direct moulding to shape the components in a single operation kept the final production cost at a competitive level. For a really advanced concept of this approach see page 129.

A polymer that promises to be less of a fire risk than materials currently used in cars and aeroplanes, for example, is a new polyester/ polyamide aromatic thermoplastic. Whereas most polymers are made up of heat-susceptible chains of carbon and hydrogen atoms, aromatic materials contain ring-shaped groups of atoms, making their molecules more stable at high temperatures. Indeed, these polymers are quite difficult to ignite and when they do burn they give off very little smoke. They are likely to replace existing polymers such as polystyrene or ABS as materials for internal fire-proof fittings in cars and aircraft.

Applications

Wear- and heat-resistant materials

These are areas in which considerable advances are being made in the application of new materials. Surface treatments based on alloying or direct ceramic coating are now being used for components that need to have a high wear resistance. Amorphous metals, in the form of thin surface-transformed regions or as thin foils are also appearing. As for high-temperature materials, the conventional nickel- and cobalt-based superalloys seem likely to give way to advanced ceramics with their superior heat-resisting and creep properties.

The properties of wear and heat resistance in materials are invariably related. For example, in machine tools wear resistance and thermal stability are highly interdependent, and gas turbine blades and fans must be able to withstand both heat and erosion. We now look at some modern approaches to these problems: surface treatment, powder metallurgy, and the use of advanced ceramics.

Surface treatments

The wear resistance of a material can be improved substantially by coating its surface. For example, it is estimated that the wear properties of cemented carbide machine tools and drill tips are improved by 15% if they have thin coatings of titanium nitride. These coatings are applied by a process called physical vapour deposition, carried out at temperatures of up to 450°C under well-controlled conditions. Layers thus deposited bond tightly with the tungsten carbide tool, providing superb wear resistance in the most demanding of machining and drilling operations.

Ceramic coatings such as zirconium oxide on cast iron are also being investigated for improving the wear and heat resistance of diesel engine parts. Already, ceramic coatings are commonly applied to components that will have to withstand wear, such as equipment for handling ore, by a plasma-spraying process. These techniques are currently being improved by introducing subsequent heat treatments, by laser for example, which make the coat-

Tomorrow's Materials

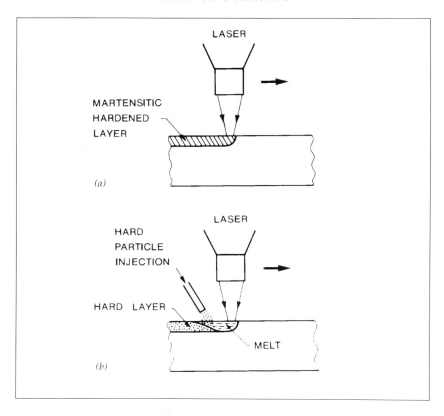

35 Surface treatment by a high-energy laser can give ordinary inexpensive metals superior corrosion or wear resistance. Two methods are (a) martensitic hardening and (b) particle impregnation

ing adhere more strongly, and also densify the porous plasma-spray coatings.

Lasers are also being used for direct hardening of high-carbon steels and other alloys. The intense heat from the laser changes the surface microstructure of the component, for example by martensitic hardening (see Figure 35(a)), or by alloying or particle impregnation in the laser-melted surface (see Figure 35(b)). The advantages of laser treatments lie in their accuracy and speed, and the fact that the process can easily be automated.

Applications

Surface treatments have the advantage that quite ordinary metals are in effect converted into quite sophisticated materials with greatly enhanced properties. Surface treated metals retain all their good properties, such as high toughness and ductility, and acquire in addition good resistance to wear and corrosion. They can even be made to look more aesthetically appealing! Surface treatment is just one of the ways in which materials can be tailored to suit practically any given design application or chemical environment.

Rapidly solidified metals

These materials are produced by various processes in which the alloy is melted down and then passed through a funnel or orifice into a reaction chamber for powder production, or directed onto a cooled wheel to produce ribbon in a process called 'melt spinning'. In the case of rapidly solidified powders the metal stream is met by a high velocity gas which breaks up the melt into tiny micron-sized droplets (see Figure 36). These droplets spheroidise due to surface tension and then cool rapidly as they fall to the bottom of the chamber. In some cases the rate of cooling is so rapid that the solidified particles barely have time to crystallise and may end up in a quasi-crystalline form - something between amorphous and fine crystalline. Since the reaction chamber contains a protective atmosphere of inert gas, contamination or oxidation of the fine powder is avoided. This is an important consideration in that the surface-to-volume ratio of the powder is very high and contamination can radically affect the properties of the sintered and densified material. An example of a single rapidly solidified powder particle of an aluminium alloy is shown in Figure 37.

The great advantage of this approach to producing metal ribbon or powders is that, even if the metal is highly alloyed, the material produced from the melt solidifies so rapidly that long range segregation or phase separation is suppressed. This means that alloys can be produced in particulate forms which would be impossible by conventional ingot casting techniques. These highly alloyed powders can then be sintered to a fully dense bulk form with little degradation due to diffusion between particles. Ideally

Tomorrow's Materials

36 Gas atomizing of a steel melt into powder particles, photographed from within the reaction chamber. Courtesy of Dr. Pelle Bilgren, Speedsteel AB, Sweden.

the powder is not exposed to air but is transported directly from the reaction chamber to appropriate containers held under atmosphere, where the powder can be cold pressed into 'green' compacts and sintered.

Two areas of application in the field of rapidly solidified metals relevant to wear and heat resistant materials are high alloy heavy duty tool steels and high alloy aluminium parts for high tempera-

Applications

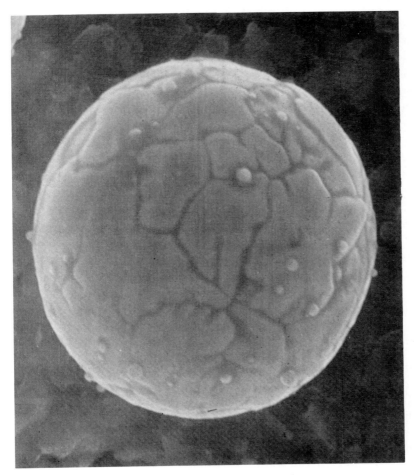

37 The microcosmology of rapid solidification showing a single Al-Mn-Cr particle. The 'canals' in this micro-Mars are microsegregation channels. The small 'moons' attached to the large satellite are powders picked up in the reaction chamber. (Scanning electron micrograph by Ping Liu, Chalmers University, Gothenburg)

(Magnification x 9000)

ture use. In the case of the tool steels, large amounts of alloying additions, such as manganese, tungsten, and molybdenum, important for producing a high density of hard intermetallic compound particles in the steel matrix, are quenched into the steel powders during rapid solidification. The material is then sintered and densified in a hot isostatic press, a special autoclave in which high pressure and high temperature can be applied simultaneously. The resultant ingots are therefore very fine grained and homogeneous and of high toughness and heat resistance -ideal properties for heavy duty machine tools such as used in milling operations for example.

Alloying constituents of rapidly solidified aluminium powders for relatively high temperature applications include manganese and chromium. Both these elements are capable of producing high density precipitate particles in the aluminium matrix. These alloys are found to be more stable at elevated temperatures than the conventional aluminium alloys based on magnesium, zinc, and copper additions. They are thus more suitable for parts likely to be exposed to high temperatures in certain aerospace or automobile applications.

Advanced ceramics

Oxygen, nitrogen, carbon, silicon, and aluminium are among the most abundant of the elements in the Earth's crust, atmosphere, and oceans (see Part III, page 147). They are also the basic constituents of what have come to be called advanced ceramics. As discussed in Part I, these materials are sintered compounds based on carbides, nitrides, and oxides, or combinations of these compounds, and are characterised by rigid and highly directional atomic bonding which provides them with great hardness and high temperature stability. Examples of some of these advanced ceramics, together with their properties and bonding types, are given in Table 7. The impressive specific modulii of ceramics have been illustrated in Table 3.

Advanced ceramics are different from conventional ceramics, mainly in the way they are made and formed by hot pressing or sintering fine powders. In this process the powders are fused together into a solid component at a temperature high enough for rapid interdiffusion of atoms to occur at particle/particle boundaries. This process can be accelerated if pressure is applied in the

Table 7 Properties of some advanced ceramics

Type	Atomic bonding	Examples	Properties
Oxides	Ionic	Al_2O_3 (sapphire) Cr_2O_3 Fe_2O_3 (hematite) MgO ZrO_2 (PSZ) $LiAl_2 SiO_6$ (glass ceramic)	Hard-wearing Good creep properties
Carbides	Less ionic Interstitial Compounds Covalent	ZrC TiC VC NbC B_4C SiC* WC	Very hard High E moduli High temperature stability Poor creep properties Used for cutting tools, abrasives, and dies

Nitrides	Covalent	BN (ambourite)
		Si_3N_4
		AlN
		Sialon†
		TiN
			Low density
			High temperature stability
			Very hard
			Good creep properties
			Used for cutting tools, gas
			    turbine wheels, nozzles,
			    and crucibles

Borides	Covalent	LaB_6 ZrB_2	Excellent conductor Used for electron microscope filaments Good creep properties

*SiC has properties more typical of nitrides.   †An alloy of Si, Al, O, and N.

same way as the powder particles of a snowball can be compacted if squeezed between the hands. The two types of ceramic are compared in Figure 38. Advanced ceramics are made by first grinding and mixing the original mineral, then purifying and filtering the powder to a fine and very uniform size, and finally densification at a high temperature. If full densification is important it is often advantageous to combine high pressure with high temperature in processes called hot pressing and hot isostatic pressing (known as HIP or 'hipping'). Alternatively, the powders may be sprayed in a plasma directly onto the surface of a metal to produce a hard coating. The resulting solid ceramic thus has an extremely fine-grained polycrystalline microstructure, almost free of residual pores and defects. It is extremely hard and cannot

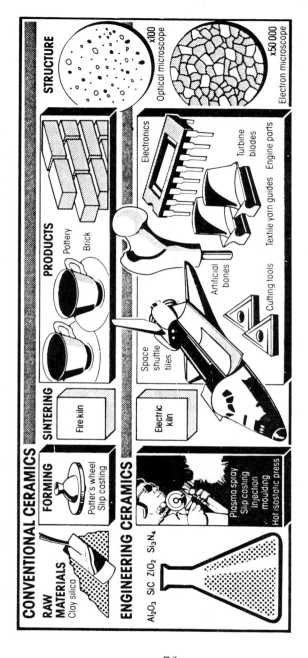

38 The essential differences between the manufacture of conventional clay ceramics and advanced technical ceramics. (Courtesy of *New Scientist*)

Applications

easily be machined or cut, except by a laser; ceramics are in general much harder than metals and will easily cut steel and glass. The current revolution in advanced ceramics was made possible by the development of techniques for producing fine powders of high purity and ways of densifying them.

Like conventional pottery, advanced ceramics are extremely strong but may break easily if hit hard with a hammer: they are not as tough or as resistant to crack growth as most metals. Figure 39 shows some typical toughnesses and bending strengths for ceramics. In fact the fracture toughness of ceramics, measured as the critical crack size for the onset of fracture, is only a twentieth of the values for the nickel-based superalloys. In other words, much more energy is needed to force a crack through a metal than through a good ceramic. In ceramic materials the crack usually follows the grain boundaries, where bonding between atoms is at its weakest. Materials scientists are now trying to improve both the sintering and the microstructure of advanced ceramics in order to increase their fracture toughness.

One of the important factors known to govern the fracture toughness of ceramics is the residual porosity- the number of pores left after sintering. Since these pores usually occur at grain boundaries, they reduce the material's resistance to grain boundary fracture. This is one of the reasons why techniques like hot pressing and hot isostatic pressing are so effective: they minimise residual porosity or remove it altogether. Another method being examined is to introduce fine fibre reinforcements into the ceramic matrix, for example silicon carbide fibres into an alumina matrix. Toughening in these systems can occur when the main crack path is diverted by debonding of the fibre from the matrix ahead of the crack (see Figure 40(b)).

Yet another way to improve the toughness in ceramics is to exploit a phenomenon known as transformation toughening. This is rather an advanced piece of materials science, and requires the use of a ceramic which is unstable in the presence of high elastic strain. This material (partially stabilised zirconia, PSZ) transforms to its normal stable crystal structure only when severely strained, and the transformation in question is 'marten-

Tomorrow's Materials

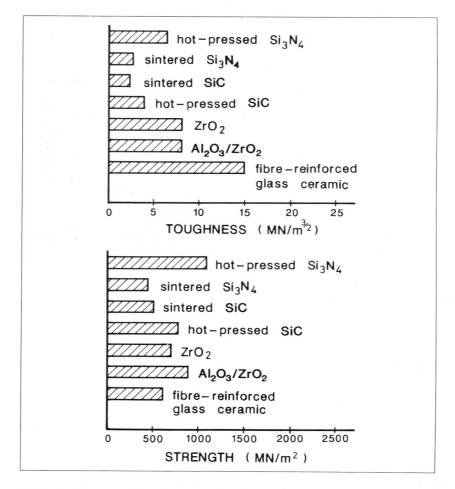

39 The toughnesses and bending strengths of some advanced ceramics. The toughnesses still lie well below those of metals

sitic', which means that it occurs virtually instantaneously. The metastable zirconia can be utilised in two ways. Very fine zirconia particles can be dispersed throughout a ceramic matrix of a different type, e.g. alumina, such that when a crack tries to grow through the alumina matrix and meets a zirconia particle, the particle transforms to its equilibrium phase and the resulting dil-

Applications

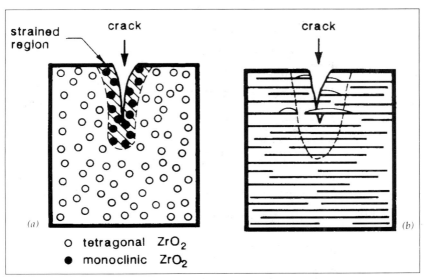

40 Two ways of making ceramics tougher. (a) Partially stabilised zirconia. the dilatation of small particles by a strain-induced crystal transformation helps to close an advancing crack. (b) Fibre reinforcement helps to divert the crack path by debonding between fibre and matrix

atation, or expansion, of the particle effectively closes the oncoming crack. The way this toughening mechanism helps to prevent crack growth in PSZ ceramics is illustrated in Figure 40(a). In an even more sophisticated approach, careful heat treatment of the zirconia causes metastable zirconia particles to precipitate within the matrix of stable zirconia. This has the advantage that a high degree of control can be exerted over the size and dispersion of the metastable precipitates. Toughened zirconia ceramics have already found use in the home as non-magnetic scissors and kitchen knives, examples of which are shown in Figure 41.

Some uses of advanced ceramics are listed in Table 8. Most advanced ceramics are produced for electrical and optical applications, and this is expected to continue for the rest of the century, at least. However, it is in applications requiring exceptional resistance to heat and creep that advanced ceramics appear to offer the most exciting opportunities for development.

Selected wonders of the world of advanced materials 3

Ceramic/ceramic composites

Ceramic/ceramic reinforced composites are remarkable for their great hardness and heat resistance combined with thermal shock resistance. Most important, their creep resistance or ability to withstand high stress at elevated temperature is the best of any material yet produced. Potential applications include turbine wheels, refractory components and very high wear components such as nozzles and dies. An important function of the fibres in these materials is to improve creep resistance. Advanced ceramics, being based on sintered powders, usually possess very small grain sizes. Since high temperature creep of these materials occurs mainly by atomic diffusion at grain boundaries, their creep resistance is dependent on grain size. The role of fibre reinforcement is thus to stiffen up the composite and prevent shape changes in components subject to high stress at elevated temperatures. An elegant example of a ceramic/ceramic composite is illustrated in the electron micrograph in the figure. The winding 'fibres' in this case are actually thin plates of boron nitride, an extremely light material of enormous strength and heat resistance. The fine grained matrix is SIALON (a silicon-aluminium-oxygen-nitrogen alloy), a ceramic of very high strength but relatively poor creep resistance. The composite is prepared by mixing the two materials and then hot pressing to obtain full density and good cohesion between the constituents. The whole is one of the most sophisticated creep-resistant composites yet contrived. It also possesses superb thermal shock resistance making it an excellent refractory material.

Applications

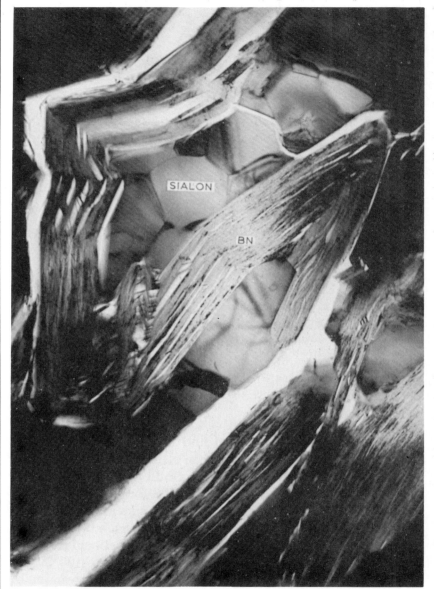

Transmission electron micrograph by William Sinclair, BHP, Melbourne

Tomorrow's Materials

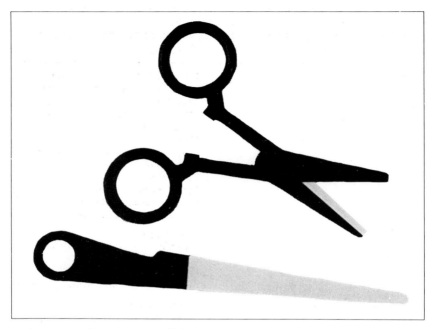

41 Two twenty-first century household appliances: a knife and a pair of scissors. They have wear-resistant blades of toughened zirconia and handles of a carbon fibre-reinforced polymer

Ceramic engines

The excellent creep resistance of ceramics makes them potentially very useful materials for diesel and automobile components, turbine blades and rotors. It has been estimated that by the year 2000, ceramics in automobile engines could have cut the world's annual energy bill by several tens of billions of dollars! This projection is based on the fact that, the hotter an engine operates, the more efficiently it runs. The use of ceramic insulation coatings, cermets, and even wholly ceramic parts in diesel engines could raise the operating temperature from about 700°C to 1100°C. This alone has the effect of improving the efficiency of the engine by almost 50%.

An advantage of using PSZ oxide coatings on engine parts is

Applications

Table 8   The main functions, properties, and applications of advanced ceramics. (After B. G. Newland, *CME*, January 1986)

Function	Property	Application
Electromagnetic	Dielectric	Integrated circuit substrate and packaging, electrical insulation
	Ferroelectric	Capacitors
	Piezoelectric	Oscillators, transducers, and spark generators
	Pyroelectric	Heat sensors
	Semiconductor	Thermistors, varistors, and heater elements
	Electrical conductor	Electrodes
	Magnetic	Ferrite magnets and recording heads
	Ionic conductor	Oxygen sensors, pH meters, and battery solid electrolyte
Optical	Optical condenser	Laser diodes
	Translucent	Envelopes for visible and infrared lamps
	Optical conductor	Optical fibres
	Optoelectric	Light valves and light memory element
Mechanical	Wear resistance	Shafts, bearings, seals, thread guides, process plant lining, and cutting tools
	High strength	Pressure sensors
	Thermo-structural	Engine components, welding nozzles, and jigs
	Low friction	Dry bearings and precision instruments
Chemical and biological	Corrosion resistance	Catalyst carriers, flow meters, and pump and valve components
	Chemical adsorption	Gas and humidity sensors
	Biological compatibility	Artificial tooth roots, bones, and joints
Thermal and nuclear	Refractoriness	Industrial furnace lining
	Infrared radiator	Thermal insulation and heaters
	Thermal conductivity	Heat exchangers and integrated circuit heat sinks
	Radiation resistance	Nuclear moderators

that its coefficient of thermal expansion is very similar to that of cast iron. Thin PSZ oxide coatings have already been used for a number of diesel engine parts including combustion chamber walls, cylinder liners and heads, piston crowns, and intake/

exhaust parts. In gas turbines, blades made entirely of silicon nitride ceramic need no internal air cooling, so they can run at higher temperatures (and hence more efficiently) than the nickel-based superalloys used at present; ceramic bearings can operate at high speed without lubricants. Silicon nitride turbine wheels and casings for turbochargers are currently undergoing trials. Figure 42(b) shows an example of a hot pressed Si_3N_4 rotor mounted in a turbocharger. Ceramic turbochargers are about 40% lighter than the conventional nickel-based alloy types currently used. This means that it takes less exhaust gas to rotate them, so the turbocharger kicks into action more quickly after the engine starts, improving the vehicle's acceleration. Ceramic matrix composites are expected to replace metallic materials in jet engines by the turn of the century. The resulting operating temperature (and hence efficiency) is expected to increase from a current maximum of 1200°C to about 1500°C.

Ceramic tools

Advanced ceramics are ideal wear-resistant materials. Even traditional wear-resistant materials like cemented carbides (tungsten carbide/ cobalt compounds) are likely to be augmented by ceramics or metal-coated ceramics in the not too distant future, particularly now that tungsten is becoming scarce. Ceramics are already widely used as refractory furnace linings, but many new applications of wear-resistant ceramics are currently being investigated: coatings for machine tools, fully ceramic machine tools, coatings for draw rolls in textile machinery, rolls and dies in metal-forming plants, and so on.

The use of ceramics in machine tools currently represents only about 2-3% of the total machine tool market, but by the end of the century it is thought that this will have increased to as much as 30%. The ceramics of interest here are Al_2O_3, Al_2O_3/TiC, sialon, and Al_2O_3/SiC fibres. Typical cutting speeds for hot isostatically pressed sialon, for example, are around 2000 rev min^{-1}, compared with 800 rev min^{-1} for TiN-coated WC and 1000 rev min^{-1} for Al_2O_3/TiC tools. Ceramic Si_3N_4 bearings and machine tools are illustrated in Figure 42(a).

Applications

(a)

(b)

42 Examples of advanced ceramic production: (a) a turbo rotor together with various machine tools and ball bearings, all of hot pressed silicon nitride; (b) the ceramic turbo rotor in position in the turbine. (Courtesy of NGK, Japan)

Bioceramics

A new and growing area for advanced ceramics is biomaterials. Traditionally, metals and polymers have been used for prosthetic or replacement devices in medical and dental surgery. In recent years, however, dozens of different ceramics have been investigated as potential implant materials. What makes them particularly suitable are their superior wear and erosion characteristics compared with other materials. For example, orthopaedic surgeons have found that the replacement of damaged bone by an advanced ceramic implant is advantageous in several respects. Not only is the high strength-to-weight ratio (the specific modulus) of considerable advantage, but advanced ceramics, during densification by a powder sintering process, can be deliberately made porous, and this enables regenerating bone to grow into and bond with the implant. Furthermore, ceramics do not corrode as do other materials used for this purpose, and they are light in weight and thus enhance body movement. The ceramics of most interest as biomaterial implants are Al_2O_3, Si_3N_4, and a complex 'bioglass' based on SiO_2. Current and potential markets for bioceramics are given in Table 9. It is thought that by the end of the century the value of this market could almost treble in value to some 10 billion dollars.

Spaceage ceramics

A novel application of heat-resistant ceramics is the well-known tiling used to protect the space shuttle during re-entry into the Earth's atmosphere. As illustrated by the electron micrograph of one of these tiles (*see* Figure 43), the tile consists of an open cellular microstructure of extremely fine coated silica fibres. The fibres are so loosely packed that the tiles consist of 95% air, making them as lightweight as cotton wool. Cellular materials are very poor conductors of heat. Trying to heat them up is rather like attempting to warm a cauliflower in the oven - the surface soon gets hot but the middle stays amazingly cool! These tiles are probably the most heat-resistant materials ever produced, being both low-density cellular and ceramic; during re-entry their outside temperature can reach 1500°C (hotter than molten steel), yet they continue to do their job.

New advanced ceramics are among the most exciting of tomorrow's materials. Although complex and expensive to produce, they are made from the most abundant of the Earth's natural materials. Materials scientists are provided with perhaps their greatest challenge in ensuring that these attractive materials, with their problems of processing and brittleness, can be fully utilised in the future. Engineering designers also need to adopt a different approach to their use since the size and effectiveness of compo-

Table 9   Current and predicted applications and costs of Bioceramic materials. (L. L. Hench, *Advanced Ceramics Materials*, January 1986)

Application	1986			1996		
	Number produced (000)	Unit cost ($)	Market ($ million)	Number produced (000)	Unit cost ($)	Market ($ million)
Total hip	300	600	180	400	700	280
Total knee	150	400	60	200	500	100
Ankles/elbows/shoulder	50	500	25	700	600	420
Finger	400	50	20	600	100	60
Fixation pins/plates	1000	50	50	1200	60	72
Tooth implants	300	500	150	1500	250	375
Periodontal treatment	200	20	4	1000	50	50
Ridge augmentation	100	20	2	2000	40	80
Ridge maintenance	250	10	2·5	15000	20	300
Mammary prostheses	400	100	40	500	120	60
Intraocular lenses	1000	400	400	1500	750	1125
Middle ear prostheses	30	150	4·5	35	200	7
Cochlear prostheses	0·5	10000	5	10	10000	100
Sutures	17500	2	35	18000	2	36
Facial augmentation	6	200	1·2	10	250	5
Myringotomy tubes	300	30	9	500	50	25
Pacemakers	190	1500	285	200	2000	400
Heart valves	40	2000	80	60	2500	150
Arterial prostheses	250	200	50	500	250	125
Ventricular assist	0·1	7500	0·75	0·5	10000	5
Vascular bypass	500	?	?	1000	?	?
Total: USA			$1403·95			$3775
(World)			($3509·87)			($9437)

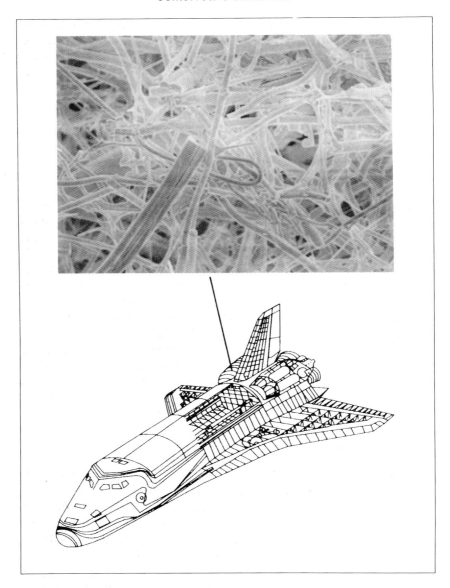

43 An electron micrograph of the open cellular microstructure of the heat-resistant tiles used on the space shuttle. Heat conduction is virtually confined to the open space between the silica fibres, and is therefore extremely low. (Micrograph by Anthony Bourdillon, University of New South Wales)

Applications

nents based on these ceramics must be far more highly optimised than those using cheaper conventional materials. In some respects quite new design philosophies are necessary. However, the potential payoff in terms of performance, weight saving, length of life, and reliability is enormous.

Optical materials

Research into these materials is currently very active, and has led to marked improvements in strength and texture, and to greatly improved coloured plate glass. Selective absorbents in glass are already widely used to filter ultraviolet light in windows in high-rise buildings and in sunglasses. In telecommunications, the use of fibreoptical materials offers huge advantages over conventional metallic materials in terms of performance (information carrying capacity) and bulk weight of the cables. These advantages are likely to increase substantially as signal losses of the optical fibres are further reduced, allowing signal transmission over hundreds of kilometres without intermediate amplification. Material for solar cells is becoming progressively cheaper and more efficient and may provide the key to cheap, 'clean' energy, even for Third World countries.

Stronger glass

Amorphous materials like glass effectively inherit the composition of the melt from which they form, so it is an easy matter to add various elements to the melt to obtain glass with whatever properties and colours are required. For window glass, additions such as sodium or potassium occupy some of the silicon sites in the glass lattice and thereby lower the binding energy between atoms in the glass molecules. This results in a material that is easier to form when hot. Additives can also be selected to help inhibit crystallisation and to improve the fluidity of the glass in its viscous state. Additives like potassium concentrate in the surface layers of glass (e.g. by ion implantation), where the presence of the large potassium atoms creates compressive stresses and so improves the glass's resistance to fracture. Alternatively, nitrogen atoms can be introduced into the glass where they partially substitute the oxygen atoms of the silica. The end-product is a glass of high through-thickness strength due to the very high covalent bonding strength of the Si-N molecules. These techniques are known as 'chemical' strengthening of the glass. Similar effects can also be achieved by thermal treatments, such as quenching the hot glass in air or a liquid. If bulk glass needs to be strengthened, nitrogen substitution or fibre reinforcement can be used.

In recent years work on the effect of additives on the properties of glass have established that, besides strength, electrical and optical properties may also be improved. In its immiscible state, the glass consists of an extremely fine interdispersed phase mixture. The way in which the two phases separate out during the liquid-to-viscous solid transition is characteristic of the precipitation of chain structures in polymerisation. The result, illustrated schematically in Figure 44, is not unlike the structure of polymer alloys. Like polymers, glass alloys can even be produced in partially crystallised form by using appropriate additives and heat treatments. Besides having better formability, these alloy glasses are found to be stronger than ordinary glass.

Alloying or doping can be used to bring about electrical conduction in silicon glass, just as doping silicon makes it semi-conducting. It seems likely that alloying glass to change its properties will continue to be an important area of development in the future, providing much stronger and more colourful glasses as well as opening up the possibility of applications in the semiconductor field.

Fibre-optics

The flow of light along fibres can be likened to the flow of water through a pipe. Advanced laser and amplifier systems have made it possible for fibres to transmit light over distances of 200 kilometres without the need for a 'boost', and such fibres are now being tested. By the end of the century the 'window thickness' of fibre optics will have reached 500 kilometres. By then the speech-carrying capacity of individual fibres will have increased enorm-

Applications

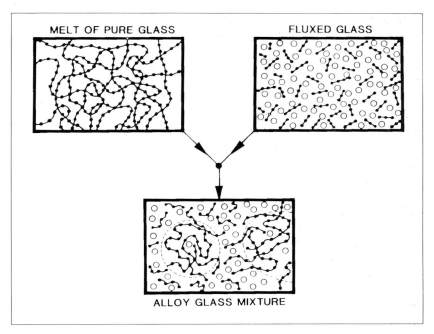

44 Tough alloy glasses are produced from phase mixtures, in much the same way that polymer blends are produced. (From R. J. Charles, 'Materials', *Scientific American*, 1967)

ously, and fibre optics will connect individual homes to international networks, bringing into sight high-resolution, worldwide TV and 'videophone' networks.

Fibre-optical telecommunication systems work as shown in Figure 45. When someone speaks into a telephone, the voice sounds are converted first into electrical signals and then into a sequence of light pulses by a laser beam. These pulses are transmitted along the fibre to the receiving end, where they are converted back to normal sound. For long-distance calls the light pulses have to be confined to a thin cross-section of very pure glass at the centre of the fibre so that there are no losses in light intensity by repeated reflection. This is achieved by making the fibres in such a way that the glass's composition, and hence refractive index, varies from relatively impure glass at the outside of the fibre 'bundle' to a fine central fibre of high purity. The change in composition over the cross-section is carefully con-

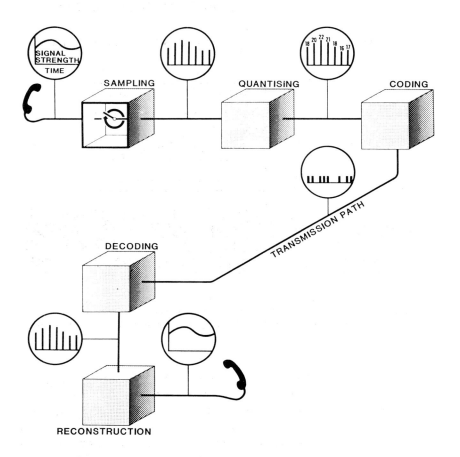

45 The application of fibre-optical systems in telecommunications. The very high frequency of the light waves allows thousands of telephone conversations to be carried simultaneously by a single fibre less than a tenth of a millimetre in diameter. (Courtesy of the Science Museum, London)

trolled by a process of vacuum deposition, so as to optimise the refractive index of the light-transmitting fibre. An example of such a graded composition fibre is shown in figure 46 on page 95. Light rays are confined to the central thousandth part of a millimetre of the fibre; light appears in effect to disobey the laws of optics by following the contours of the fibre as it winds its way

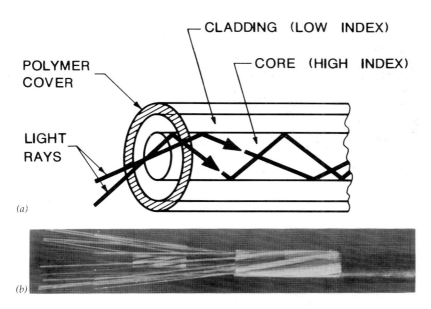

46 (a) How light waves bounce along a fibre enclosed in a sleeve of high refractive index. The secret of modern fibre optics is to reduce these reflections to a minimum, thereby increasing the efficiency of transmission. (b) A fibre-optical cable

from one location to another. The loss of light-transmitting efficiency suffered by modern fibres amounts to only 0.2 decibels per kilometre. This means that light can travel 20 km along the fibre before losing half of its original intensity. This efficiency will certainly be improved in the future, and 0.1 decibels per kilometre should be achievable even with present-day technology.

An important advantage of glass fibres over conventional copper conductors in telecommunications is the extremely small diameter of the fibres compared to their metallic counterpart, together with their far greater load-carrying capacity (see Figure 47). Typically glass fibres have a diameter of less than a tenth of a millimetre, so even when clothed in their polymer jackets, many more fibres can be fitted into a cable of a given size than can copper wires. Furthermore, unlike metallic cables, fibre-optical carriers are not affected by electromagnetic disturbances. This has, for

Selected wonders of the world of advanced materials 4

Optical fibres

These materials are remarkable because of a sophisticated processing route which provides fibres with enormous information-carrying capacity but very small losses in light intensity. Thin layers of different compositions are deposited on the inside of a silica tube using a deposition process from flowing gas. A computer monitors the composition of the gas so as to continuously modify the deposited layers of material. Following this, the tube is hot drawn into long, solid fibres of about a tenth of a millimetre diameter. In this form the graded change in composition produced by the deposition process effectively controls the refractive index of the glass fibre, such that the transmitted light beam is confined to the inner thousandth of a millimetre of the fibre. In this way, the loss of light intensity due to 'bouncing' within the fibre are reduced almost to zero. More recently, fluoride fibres have been produced with such low losses that light can be transmitted over hundreds of kilometres without need of reamplification. The figure illustrates a silica fibre, and its corresponding graded composition is shown in the form of concentric rings within the cross-section of the fibre.

Applications

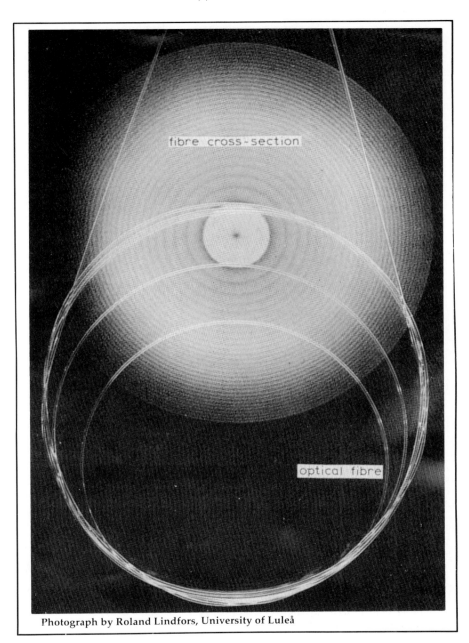

Photograph by Roland Lindfors, University of Luleå

Tomorrow's Materials

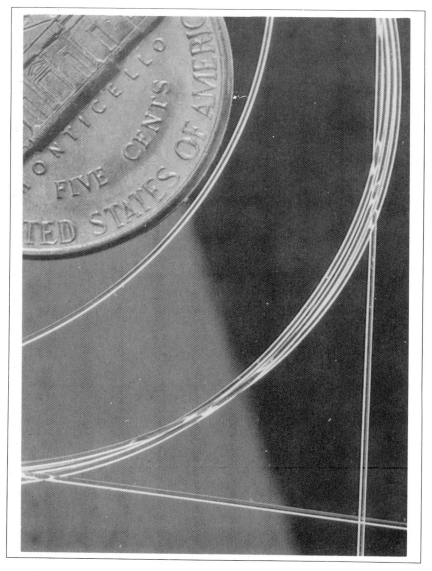

47 A single optical fibre as illustrated can carry simultaneously some 8000 telephone conversations or 60 TV video transmissions. These not-so-delicate silica threads hold the key to worldwide TV and vision telephone networks by the turn of the century. (Photograph by Roland Lindfors, University of Luleå)

instance, opened up the possibility of suspending the cables over railway lines. Already in the USA networks of fibre-optical cables are linking individual homes to the national TV networks, and similar fibre-optical networks for 'videophones' are projected. With plans for transoceanic fibre-optical cables progressing well, these advanced telecommunication systems will shortly span the world.

Recent research suggests that scientists are not satisfied with the very low optical losses of glass fibres, and are already investigating other new materials with the object of improving light-transmitting efficiency still further. For example, fluoride fibres, with complex compositions at the core, have been produced and found to exhibit only a tenth of the losses of the best silica fibres.

Solar cells

Like fibre optics, solar energy conversion is likely to become a very important technology. It promises to be a cheap and 'clean' energy source, particularly suitable for Third World countries where other energy sources, based for example on oil or atomic power, are expensive, or undesirable.

One of the best solar cell materials is crystalline silicon, the material used for example in space probes and household solar converters. Crystalline silicon has a higher conversion efficiency than amorphous silicon, although it is an expensive material. While high costs may be acceptable for sophisticated space applications and luxury homes, the cost of large ground-based solar energy installations using crystalline silicon would be prohibitively high. A cheaper alternative material for large-scale solar cells may be amorphous silicon. It can be cheaply produced in the form of very thin coatings evaporated on to a suitable substrate material. Even very thin evaporated films, of the order of a thousandth of a millimetre thick, are able to convert solar energy fairly efficiently.

Current research is aimed at improving the efficiency of amorphous thin-film silicon cells, for example by employing layered structures rather like those found in modern fibre optics. In this way, it is envisaged that giant solar energy installations receiving sunlight reflected down from orbiting satellites will

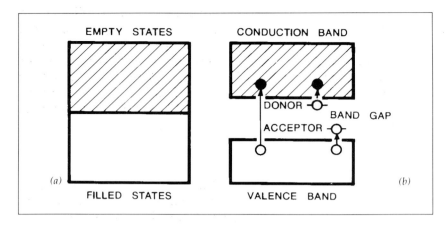

48 The energy bands of (a) a metal and (b) a semiconductor. An input of energy to the semiconductor in the form of light or heat causes a valance (free) electron to jump up to the conduction band. The role of graded impurities, or dopants, is to provide 'carriers' with energies within the 'forbidden' gap. Some dopants donate electrons to the conduction band, while others accept electrons, creating 'holes' in the valence band. Doped silicon chips are based on this principle

meet the energy needs of whole communities and even industrial plants. Other ambitious projects being considered include the possibility of achieving solar energy conversion in space, where the process would be much more efficient. The energy could then be beamed to a receiving station on Earth by using microwave transmission.

Electronic and magnetic materials

Just as steel changed the world during the industrial revolution, so the semiconductor has had a similar impact in the 1970s and 1980s. 'Silicon Valleys' are now found in many places other than California, and the importance of silicon to our chip-shaped world is likely to continue for some years to come. However, new semiconducting materials are appearing on the horizon, some of

Applications

which will certainly soon begin to steal some of the limelight from the ubiquitous silicon. A milestone in this field has been the discovery of the 'high-temperature' superconductors based on copper oxide and capable of operating well above liquid air temperatures. These remarkable ceramic materials herald a whole new generation of devices- miniature high-speed computers, super-magnets, levitating trains, and compact nuclear magnetic resonance scanners for detecting tumours.

A chip-shaped world

Today's semiconductors are based mainly on crystalline materials, for example silicon or gallium arsenide, doped with boron, arsenic, or phosphorous. As illustrated in Figure 48, the role of these dopants is to pass electrons to the host material or accept electrons from it, causing it to become conducting.

As shown in Figure 48(a), metals are conducting without the need for alloying additions or dopants because of its partially filled conduction band, in which electrons jump easily from the filled to the empty parts of the band, where they are free to move from ion to ion. In Figure 48 the metallic conduction band is represented by both empty and filled states. Because there is no band gap, very small changes are needed for electrons to move from the filled to the empty state. With semiconductors, however, an input of energy in the form of light or heat is needed to cause electrons to jump over the band gap and bring about conduction. Silicon semiconductors have quite a wide band gap, and the energy required to bring about conduction is therefore quite high. This causes the chip to warm up, and in large computers like the Cray II efficient cooling systems are essential. Other semiconductor materials such as gallium arsenide have smaller band gaps, so less energy is needed to activate them. Materials such as these will provide more efficient semiconductors in the near future.

The technological importance of semiconductors is that the contact between a semiconductor and a metal (a full conductor) acts as a rectifier, enabling electric current to pass more easily in one direction than the other. In computing, for example, the ability to provide low-energy switches, in which small rectifying junctions are used to turn current on and off extremely rapidly, is vital to the various operations computers are required to carry out.

A problem with crystalline materials such as silicon or gallium arsenide which are to be doped is that they must be in an extremely pure form, otherwise unwanted defects such as impurity atoms or even dislocations tend to disturb or interrupt the flow of electrons. It is worth taking a closer look at the materials science behind semiconductor production, particularly silicon chips.

Silicon chips

The raw material for silicon is quartz or silica, which is melted and refined to produce metallic polycrystalline silicon. This metal is then melted in a crucible, and a 'seed' crystal is used to grow a single crystal of silicon . This must have a minimum purity of 99.999 999 9% . The next step is to cut the crystal into thin wafers, about a quarter of a millimetre thick, whose surfaces are ground and polished to as smooth a finish as possible, in order to remove as many defects as possible. Depending on their size, about 500 chips are laid side by side on a wafer. Each chip contains about a million components or junctions. These junctions consist of negative and positive charge carriers, in which electron 'holes' are positive (*p* types) and excess electrons are negative (*n* types) (*see* Figure 48). Used together, these junctions create all manner of electronic components such as transistors, diodes, and capacitors. The *p* and *n* junctions are produced by a highly selective doping procedure in which, for example, boron atoms in the silicon create the electron holes and phosphorus produces the excess electrons. As illustrated in Figure 49, this is achieved by coating the wafers with silicon dioxide, and then with a photosensitive material. The circuitry pattern required is then exposed photographically onto the photosensitive coating, and permanently imprinted by hardening the layer. The unhardened layer is finally removed by an acid leaving the pure silicon base unaffected. Doping is then carried out by exposure to an atmosphere of phosphorus or boron at a temperature sufficiently high to cause the atoms to diffuse from the atmosphere through the 'windows' in the silicon dioxide layer produced by the masking process. The production of silicon

Applications

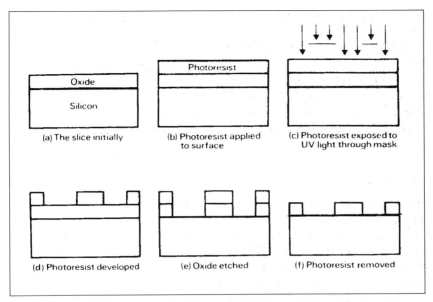

49 A schematical illustration of the process of producing a silicon chip. (Courtesy of the Science Museum, London)

chips is a very demanding one, and nowadays up to about 50% of all chips produced are rejected because of the presence of small atom-sized defects in the silicon lattice, as well as other problems.

After silicon

As we have seen, a problem with silicon chips is that the energy needed to bring about the flow of electrons causes the chip to heat up, and that as a result in a large computer measures must be taken to avoid overheating. A better material in this respect is gallium arsenide (GaAs), which has the advantage that less energy is required to induce flow, so there is less of a heating effect than in doped silicon. As noted in conjunction with Figure 48, the band gap of a semiconductor defines the energy needed to raise electrons into conduction states or, conversely, the energy given out when electrons move out of conduction states. In compound semiconductors like GaAs this energy is usually in the form of light waves. Devices which can exploit this type of two-way

Tomorrow's Materials

conversion of energy between light waves and electric currents include lasers and solar cells. An example of the way gallium arsenide laser devices are constructed is shown in Figure 50.

Gallium arsenide has a much higher electron mobility than silicon, and as such it is a better material than silicon for use in low-energy high-frequency applications such as fibre-optical lines, remote control devices for TV, and car telephones. A disadvantage, for the materials scientist, is that gallium arsenide is more difficult to produce than silicon in a form that is pure and free of defects. An alternative approach is to evaporate thin, defect free, films of GaAs onto a silicon substrate material. Chips of this composite material have been produced in which islands of GaAs are employed as ultrasensitive optical switches within the silicon semiconductor circuit, thus combining the best qualities of both materials. It is predicted that the world market for gallium arsenide semiconductor devices will increase from the present-day 200 million dollars to around 4 billion dollars by the turn of the century.

Gallium arsenide is not of course the only new potential semiconductor material being examined. The glass-based selenium-germanium alloys are an interesting example of how amorphous materials can be used as semiconductor materials. When a pulse of electrical energy is applied to this material, the glass heats up to a temperature just above the glass transition temperature, causing the glass to crystallise locally in the form of conducting fibres that span the amorphous matrix film. These fibres are so fine that they contain no defects, so conduction along them is extremely efficient. The applied electrical pulse thus has the effect of rapidly increasing the conductivity of the material. When a second pulse is applied the crystalline fibres melt and then cool so rapidly that they revert to the glass or amorphous structure, with the result that the device returns to its former low conductivity state.

These amorphous semiconductor materials are currently being considered for use in computer memories as erasable 'read-only' devices. It is also possible that the energy pulses needed to bring about the crystallising transition could be supplied by a laser. It may therefore be feasible in the near future to use these simple

Applications

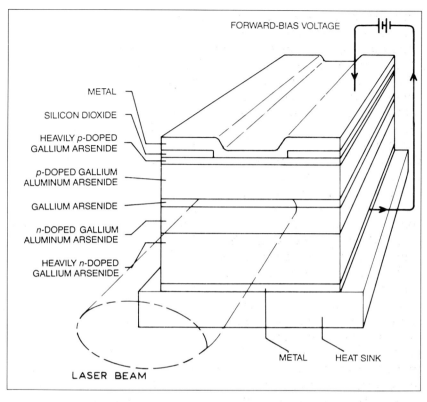

50 A gallium arsenide heterojunction laser device. The various material layers, deposited by physical vapour deposition, effectively confine emitted light to a narrow slit. (J. M. Rowell, *Scientific American*, Oct. 1986)

types of semiconductor in transmitter/receiver devices in fibre-optical telecommunication systems.

Conducting polymers and plastic chips

Today, semiconducting polymers are used in a variety of applications including electronic and electrical components such as switch contacts, resistors, and electrodes, and as guards against lightning and electrical discharge. Apart from electrical conductivity, the associated property of good thermal conductivity is

Tomorrow's Materials

also useful for obtaining rapid through-heating of parts produced in extrusion and forming operations.

Two ways have so far been used to make polymers electrically and thermally conducting. One is simply to embed a fine dispersion of conductive fibres or particles into the plastic; the other is to paint a conductive coating on the surface of the part. However, a more fundamental approach is to modify the conduction bands of the polymer molecules by alloying or doping.

The use of doping to increase conductivity is a recent innovation of scientists and provides the key to the development of the plastic chip. A dopant such as arsenic pentafluoride is added to a suitable polymer such as polyacetylene. As in silicon semiconductors, the effect of doping is to modify the conduction bands of the polymer molecules so as to bring about electron acceptor/donor manipulations. For example, by exposing polymers such as polyacetylene to an iodine vapour the polymer is made to give up electrons to the vapour phase, increasing the mobility of electrons in the polymer; in other words it becomes a conductor. Another approach is to dope graphite fibres with iodine or bromine. This causes ions to form which help transfer charged particles along the fibres. However, there are problems with producing conducting polymers in these ways. Doped polymers are not very stable in air, and the dopant quickly leaks away. A conducting polymer that is both cheap to produce and stable in air is needed if plastic chips are to be a realistic proposition. Such a polymer does not exist today, though given the present level of research activity in this area the cheap plastic chip will probably appear by the turn of the century.

In summary, silicon semiconductor chips may eventually be displaced, or maybe even superseded, by chips made from gallium arsenide or even plastic. On this basis, tomorrow's computers and electronic devices will operate more rapidly, need less energy, and require less cooling than the machines we use today. Even the elements connecting individual chips will be more efficient in future, as they will probably be based on optical fibres, or even superconductors, which not only transmit faster than normal metal conductors, but also will not impose the high capacitive loads that slow down and heat up today's computers.

Applications

Magnets

Magnets are to be found all around the home: they energise the electrical motors in vacuum cleaners and refrigerators, they help reproduce sound and visual images in stereo and video equipment and store information in computers, and they keep cupboard doors closed. Indeed, man has used magnets - or at least known of them - for at least 3000 years, thanks to the abundance of the magnetic mineral magnetite, commonly known as lodestone. By the sixteenth century it had been demonstrated that if freshly smelted iron is drawn in the shape of a bar, it becomes magnetic.

From the technological point of view an ideal magnetic material should have one of two properties: it should be either 'soft', meaning easy to magnetise or demagnetise, or 'hard', meaning that it remains magnetic or is a permanent magnet. Whether a magnet is hard or soft is decided by whether it remains magnetic after an externally applied electrical field has been removed. This degree of magnetisation, in terms of the magnet's remanence and coercivity, is illustrated in Figure 51, and is often referred to as the magnet's 'hysterisis' after the Greek word meaning 'to lag'. Generators, electric motors, and transformers operate at their maximum efficiency if magnetisation does not remain after the external field has dropped to zero, and this requires a soft magnet of low remanence with the type of hysterisis loop shown in Figure 51(a). On the other hand, permanent magnets, used in refrigerators, doorcatches, and earphones, need to have a higher remanence with a hysterisis loop of the kind shown in Figure 51(b).

The best traditional permanent magnetic materials are based on 'ferrites'. Over the years, however, alloy development has resulted in substantially stronger (harder) magnetic materials than the ferrites, as illustrated in Figure 52. In more recent years, hard magnets have been made from sophisticated alloys that can be produced in a particulate form, such that each particle effectively represents a single magnetic 'domain'. These individual magnetic particles can then be aligned under an applied electrical field and subsequently sintered or mixed with a suitable binding material to make very powerful magnets indeed. In another approach, materials such as an iron-boron alloy, with additions of silicon or neodynium, are formed into amorphous ribbons by

Tomorrow's Materials

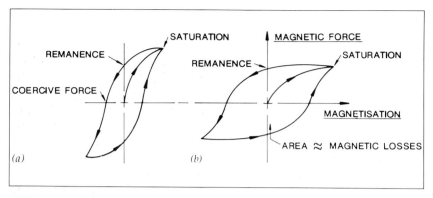

51 The magnetic loss or 'hysteresis' curves for (a) a soft magnet and (b) a hard magnet. The significant difference is that soft magnets easily lose their magnetisation, whereas hard magnets remain magnetic. These characteristics determine their uses

very rapid solidification. These materials show extremely low hysteresis losses when magnetised, which means that the magnet is very soft and remagnetisation can occur any number of times with low losses. Such materials are now being used as laminated core material in transformers and electric motors, and for audio tape recording heads.

Other applications of magnetic materials call for properties that lie halfway between hard and soft magnets. The magnetic memory element in a digital computer, for example, must be hard enough to retain its forward or reverse magnetisation ('zero' and 'one' states in the computer's binary system), yet soft enough to switch states cleanly and rapidly when a small external load is applied during read-in or read-out procedures. For this purpose, tailor-made materials are being produced based on 'bubble memory' characteristics which change their properties in a supersensitive way, not unlike some of the functions of the human brain. However the future may again be in the hands of polymer chemists. Soviet researchers, for example, have recently produced 'ferromagnets' based entirely on polymers (polydiacetylene), with potential for a wide range of new magnetic materials.

Applications

Superconductors

If certain materials are cooled to temperatures of some 200°C below the freezing point of ice, something almost miraculous occurs: they become superconducting - able to conduct electricity with virtually no loss of energy in the process. Normal electrical transmission lines, for example, have energy losses of 15-20%, so that if only superconducting cables could be used at reasonably high temperatures ('high' here means above the temperature of liquid air, -196°C), this would be a major breakthrough with great commercial significance in power transmission and many other applications. The liquid air temperature, -196°C, may sound cold - and hence expensive to create - to some people (it is about three times lower than the mid-winter temperature at the South Pole), but in fact liquid air is relatively cheap to manufacture, costing about the same as beer. Recently, such a breakthrough in 'high-temperature' superconductivity has been made, with the discovery that certain ceramic materials based on copper oxide are superconducting at temperatures well above the freezing point of air. Quite suddenly, the realm of superconducting transmission lines, miniature high-speed computers, and supermagnets is on our doorstep.

The discovery of superconductivity resulted from experiments by Heike Kamerlingh Onnes, a Dutch scientist at the University of Leiden in 1911, who found that the electrical resistance of mercury decreased as expected on cooling, but on reaching a temperature of 415 K (about -269°C) its resistivity suddenly vanished abruptly and completely. As illustrated in Figure 53, since that time - and despite much research - little progress was achieved in finding a material that loses its resistivity above liquid air temperature. Until 1986 the record transition temperature, set by a metallic niobium-germanium compound, was only -250°C. The breakthrough came about in that year when researchers at the IBM laboratories in Zurich achieved a superconducting temperature of -237°C in a copper oxide compound containing minor additions of lanthanum and strontium. That superconductivity should be achieved in a ceramic, the most traditional of insulating materials, was amazing in itself and immediately provoked a new wave of interest in the subject.

Selected wonders of the world of advanced materials 5

'Warm' superconducting oxides

These materials are remarkable in that they possess zero resistance to the flow of electrical current at a relatively 'warm' temperature, above that of liquid air. The most reliable and technologically interesting of these oxides is based on a copper oxide containing additions of yttrium and barium. Oxides or carbonates of the constituents are ball milled into extremely fine particles, cold pressed, and then fired (or sintered) at around 925 °C. By cooling the partially sintered compact slowly, the lattice structure takes up oxygen and an unusual distorted perovskite-type single phase oxide is formed with the composition $YBa_2Cu_3O_7$. This material is metallic in appearance and properties and has in fact an ambient temperature conductivity almost as low as that of copper. Most remarkable of all, however, is that if it is cooled to a temperature of about 90K (*ca.* 23 °C above the temperature of liquid air) it becomes a superconductor. Since it is easy (and inexpensive) to produce liquid air, these superconductors have enormous potential use, for example, as zero-loss power transmission lines, super-strength magnetic materials, and materials for levitating trains. The figure demonstrates the fine microstructure of this superconducting oxide. The striations across the elongated crystals are thin 'twins' or faults caused by the transformation from the high temperature tetragonal (non-superconducting) phase to the low temperature orthorhombic (superconducting) phase.

Applications

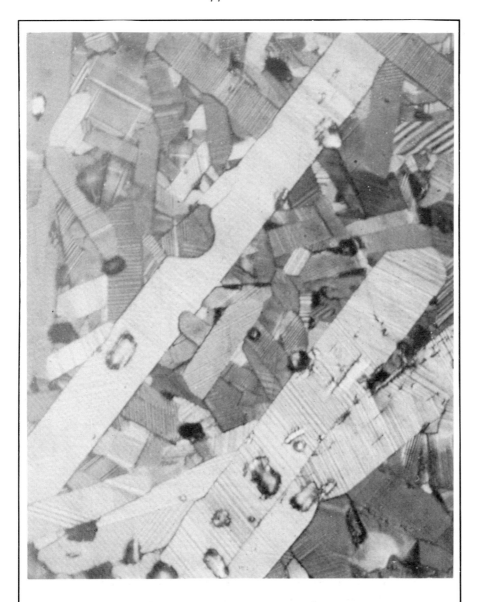

Light optical micrograph by J. P. Zhou, University of New SouthWales
(Magnification x 325)

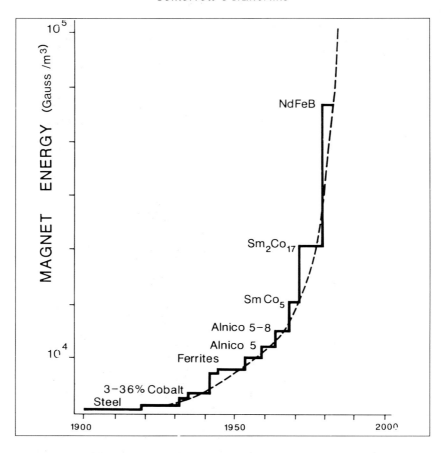

52 This graph illustrates how the strength (as measured by maximum magnetic energy) of permanent magnets has increased with the development of new materials. The powerful new magnets will be used in a wide variety of applications, such as lightweight headphones and car telephones

We still do not understand why these or any other materials are superconducting, so obviously it is difficult to say why ceramic materials should be better superconductors than others. The phenomenon occurs when the normal process of scattering between electrons and the compound's lattice are suddenly removed. The materials science of this is complex and interesting. Superconduction is thought to be associated with superfine-

Applications

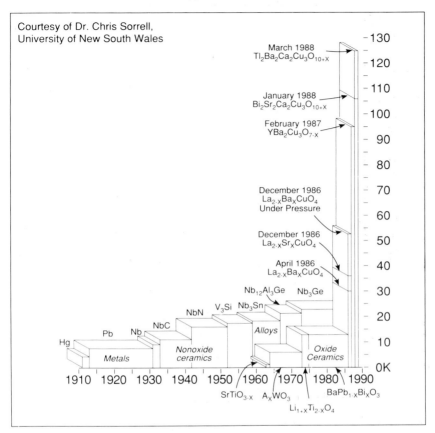

53 The development of superconducting materials over the years, in terms of their maximum temperature of superconduction. The temperature of liquid air is 77K.

layered structures of just one or two atoms thick, which somehow distort the lattice and amend its valency bands sufficiently to bring about zero-resistance free 'tunnels' through which electrons or electron pairs move unimpeded. The difference between superconducting, normal conducting, and insulating behaviour is illustrated in Figure 54.

The electron collisions (Figure 54) which occur in normal conduction are due to interactions between the moving free (valance) electrons and the phonons. Phonons are quantised wave phenomena brought about by the lattice vibrations of the atomic structure.

Tomorrow's Materials

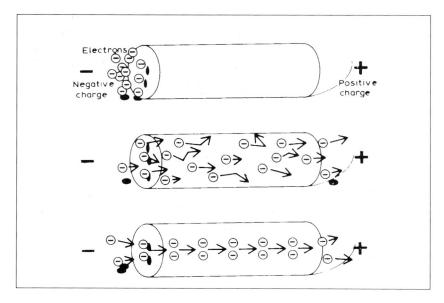

54 The flow of electrons in an insulator, a normal conductor, and a superconductor. (From *Time*, 11 May 1987)

It is conjectured that in superconduction there are no collisions between the electrons and phonons; instead, the electron pairs that form up move in unison with the lattice vibrations so as to assist the electrons along their paths. Normal electron conduction could be compared with paddling out to sea on a surf board, while superconduction is like riding the board back on the incoming waves.

In the case of the new copper oxide based superconductors, it is found that addition of rare earth elements such as lanthanum, strontium, scandium, yttrium, bismuth or thallium to copper and barium oxide completely changes the normally insulating properties of the oxide to produce a superconductor. Indeed these materials are fairly good conductors even at room temperature, with properties and appearance not unlike that of a typical metal. The most successful and stable of these superconducting oxides has the atomic composition $YBa_2Cu_3O_7$. This compound exhibits zero resistivity and superconducting properties below a temperature of -175°C, some 22°C *above* the temperature of liquid nitrogen

Applications

(*see* Figure 53). Materials scientists are currently experimenting with compositional modifications to further increase the zero resistivity temperature. In terms of overall stability and repeatability during temperature cycling, the $YBa_2Cu_3O_7$ material is currently the best candidate as far as technological application is concerned. However, new and possibly better materials are on the way. For example, oxides based on bismuth or thallium have T_c's up to 125K, or 44 °C above that of liquid air.

An example of the atomic configuration of an yttrium-based compound is shown in Figure 55. It is seen that the structure is layered in the sense that it contains conducting planes dense in copper and oxygen atoms (shown shaded), separated by non-conducting planes corresponding to the positions of the yttrium and barium atoms. Superconduction occurs along these copper and oxygen electron-rich planes as indicated by the arrows. Indeed, all of the superconducting oxides exhibit this layered form of structure.

Of new materials being investigated, oxides based on bismuth appear to hold most promise. New materials of a Bi-Sr-Ca-Cu-Oxide have now been produced in Japan, Sweden and Australia in the form of thin, silver-coated tapes, giving a T_c of 103 Kelvin and a current density as high as 12 000 Amps/cm^2. The secret in this case is to prepare the material in such a form that it contains a fine dispersion of flux-pinning centres made up of tiny calcium-copper-oxide particles. Flux pinning, even in conventional (low-T_c) superconductors, was found to be essential if high currents in a high magnetic field were to be achieved. Having the superconducting material in the form of these tapes, sealed in silver, has a number of advantages: the oxide is protected from the environment; the tapes are fairly flexible; they easily re-absorb oxygen (lost during processing) through the thin silver coating; they can conveniently be stacked in layers within a 'bulk' conductor; the silver can act as a good (standby) conductor if overheating occurs and the material stops being superconducting. Such tapes are now being considered as potential conductors for under-city power transmission cables and for super-magnets.

Making power cables out of these ceramic compounds will certainly be a task to tax the ingenuity of materials scientists. For most applications the superconducting wire ought ideally to be a

Tomorrow's Materials

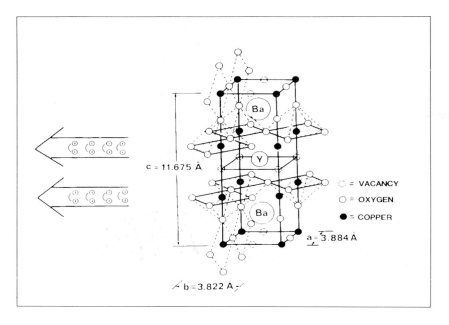

55 A unit cell of the superconducting ceramic YBa$_2$Cu$_3$O$_7$. The structure is orthorhombic in form and the 'vacancies' refer to 'missing' oxygen atoms. Superconducting planes are shown shaded and the direction of superconduction is indicated by arrows

good 'normal' conductor at ambient temperature, otherwise there is the risk of the cable being destroyed by sudden heating as it warms up above the superconducting transition temperature. To achieve this it may be necessary to integrate the superconducting compound with, for example, normal copper cable, unless compounds retaining their superconductivity up to ambient temperatures can be produced.

Potential applications of superconductors are many. Power transmission lines have already been mentioned. It would be easier to lay these underground, under cities for example, than it would normal cables, which tend to have high energy losses and suffer from heating problems. Schemes for cooling underground superconducting cables have already been advanced, including passing cold gases over the superconducting lines within sealed covers.

Applications

56 'Levitation' - the magnetic field of a magnet is repelled by a superconductor causing the magnet to hover above the superconductor. This is known as the Meissner effect

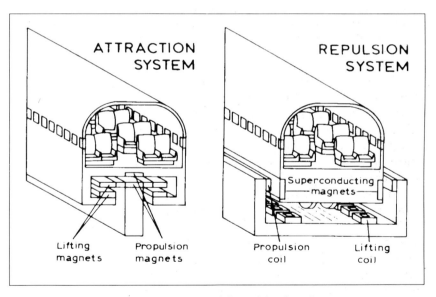

57 A train is made to levitate by the powerful repulsive force between superconducting magnets in the train and magnetic 'railway lines'. (From *Time*, 11 May 1987)

Tomorrow's Materials

In computers, superconducting switches can flip at least as fast as gallium arsenide chips, and when used as a circuit component a superconductor can communicate with other components faster than normal conductors. In addition, superconductors dissipate practically no heat and chips can be packed tightly together, making further miniaturisation possible. In medicine, metal superconductors operating at liquid helium temperatures (-269°C) are already used in nuclear magnetic resonance scanners. These scanners produce images showing the concentration of certain atoms in the body, and hence provide information on the function and condition of the body. For instance, they can help to locate the presence of small tumours. Very high magnetic fields are normally needed for this technique, requiring superconducting magnets cooled to liquid helium temperature. Superconducting oxide magnets operating at higher temperatures need only be cooled by liquid air or carbon dioxide. With the help of the new superconducting ceramics, it is quite conceivable that a nuclear magnetic resonance scanner of much simpler construction could well be in every specialist's consulting room by the turn of the century.

One of the unique characteristics of superconducting materials is that they repel ordinary magnetic fields. This gives rise to a kind of levitation (*see* Figure 56) in which, for example, superconducting materials could be used to keep trains clear of magnetised rails, as illustrated in Figure 57.

There are many other ways in which high-temperature superconductors may affect our lives. As with the development and application of other advanced materials, the ways in which these remarkable materials are put to the service of mankind is limited only by the imagination and conscience of the citizens of tomorrow's world.

Applications

Sporting Materials

The latest design concepts and the most advanced materials are nowadays used in sports equipment. Concentrating here on sports in which human endeavour, rather than mechanical power, is the driving force, the latest trends in design and materials is looked at in jogging shoes; tennis and squash racquets; bicycles; surfboards; canoes and boats; and skis. It is found that more emphasis is now being placed on ensuring freedom from injury than before, and that at best, sports equipment in future will be optimised to suit the limit of the enthusiast's ability, and pocket.

Sports Shoes

In an age when jogging has become a sort of cult, when a whole cross-section of the population is to be observed puffing and blowing along the sides of the road, when more people than ever before are taking up active sports, the need has arisen for sports shoes that can both accommodate the punishment of long distance running and other sports, as well as help protect the wearers from injury and fatigue. Perhaps, more than all this, sports shoes have become a cult in themselves. They are not only worn for sport or running, but as daily footwear by large sections of the population. Their design, incorporating advanced engineering polymers and composites, is in many respects revolutionary.

Human feet come in all sizes, shapes and forms. They can often be grossly misshaped and are nearly always sweaty! Yet anatomically feet are extraordinarily complex constructions (Figure 58) and as such highly susceptible to injury when misused or overloaded. This may well excite the interest of chiropodists and chiropractors, but sports shoe manufacturers have the unenviable task of constructing shoes that protect and support these anatomical extremities under the most demanding conditions of pounding, sweating and foul weather; no small feat, indeed! This is where the wonders of materials sciences have come to the rescue, substituting the 1000 year-old traditional shoe material of leather soles and uppers by a delicately designed, super-lightweight, multi-polymer composite.

Tomorrow's Materials

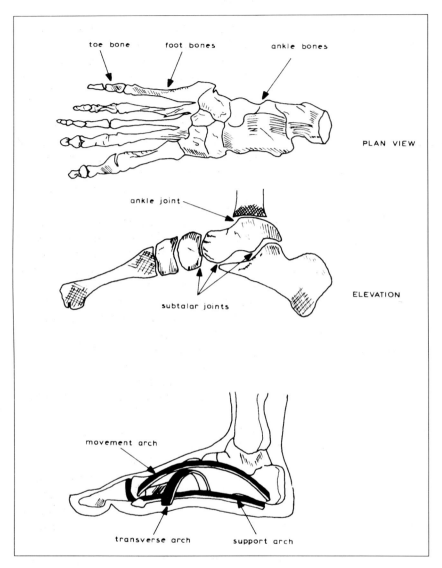

58 (a) Side view and plan projections of the bones of a human foot (b) The foot represented in terms of its three support arches. Note how modern sports shoe design tries to compensate against overstressing these arches (see Fig. 60, p.122). Adapted from Rolf Wirhed's "Athletic Ability and the Anatomy of Motion" (Wolfe Medical Publications Ltd., 1984)

Applications

The main problem, as seen from Figure 58, is that running subjects the arch of the foot to a high stress. The various ligaments of the foot and ankle, together with the wedge-shapedconstruction of the foot's bones and the muscles, all give the foot its elasticity and ability to absorb stress under normal use. Many runners, however, have a lopsided gait, producing excessive wear on the outside heel of the shoe. In addition, most running shoes wear badly beneath the big toe. These characteristics of running need to be compensated in the design of shoes and by the use of appropriate dampening materials in the shoe.

The main requirements of sports shoes are that the large and sudden pressures acting on the feet (and limbs) during any given activity are minimised. In addition, exaggerated sideways glide, and twisting of the foot needs to be avoided or compensated if possible. Feet morphologies basically fall within two extremes, with a range of permutations in between. At one extreme is the rigid, high-arched foot (not so common) that does not spread much. For this foot, cushioning or shock absorption is important. At the other extreme is the flat foot which easily spreads, and this is the type of foot that needs sidewise support.

Yet another property of sports shoes is their energy of rebound. This is illustrated in Figure 59 showing the degree to which energy 'recovery' can be achieved. This recoverable rebound energy, represented by curve B in the figure, is the property of the shoe that helps propel the wearer along. In most sports shoes, this energy return is as much as 65% of the total energy absorbed (the claims made by certain manufacturers that their shoes are 'bouncier' than others, is largely unfounded!). However, shoes that emphasise rebound in their design philosophy are not really catching on. Most sportspeople simply want to keep fit, not break records. Thus the overriding philosophy today in the design of sports shoes is to take care of the absorbed energy, or 'cushioning effect', thereby holding injury to a minimum.

To understand how the cushioning and rebounding energies differ in the case of running and jogging, consider how the shoe experiences the complete loading cycle. Whether walking or running, we normally meet the ground with the heel, creating a brief but jolting force of up to 30-40 G! After the heel takes the initial shock, the load transfers to the ball of the foot and there

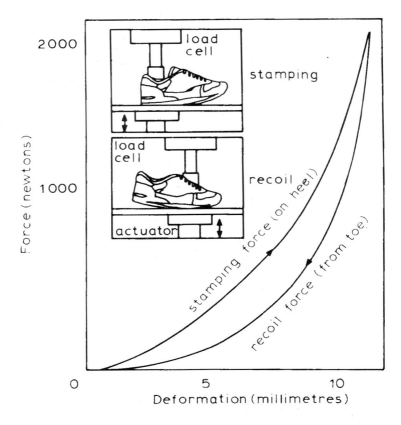

59 Measured stamping and recoil forces on a sports-shoe. The shaded area is a representation of recoil efficiency. After R. McNeill Alexander and M. Bennett, *New Scientist*, 15 July 1989, p.45

reaches a maximum. At this stage, the front part of the sole of the shoe is highly compressed, and thus a maximum of energy of rebound occurs as the foot leaves the ground again. The force maximum on the shoe has been measured to be as much as 2-3 times the body weight (say, 2000N).

In order to accommodate countless loading cycles like this, and yet continue to function well, sports shoes have to be fairly complex constructions, consisting of at least five or six main parts (see Figure 60, p.123). The shoe thus consists of: an outer sole, an

Applications

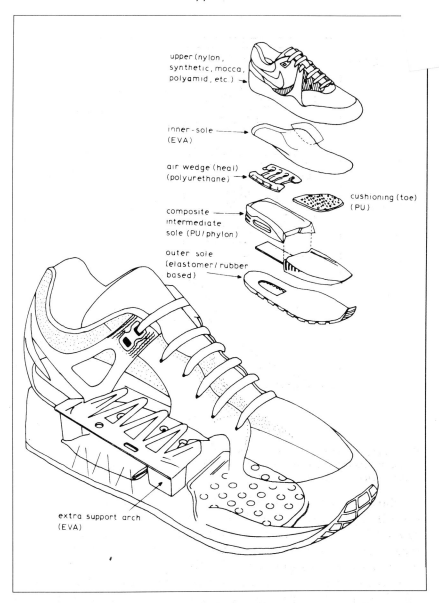

60 Illustration to show the make-up of a modern sports-shoe. Note the support arch both cushions and compensates lateral glide. (Collected and assembled from a Nike brochure)

inner sole, an intermediate sole, plus cushioning and supporting components at the heel, the arch and the ball of the foot. In addition, there is usually a cup placed at the heel for extra support. Cushioning is usually provided by an ethyl vinyl acetate (EVA) and polyurethane foam. Densities of these materials can be varied to suit the individual's anatomical needs. The use of a composite of at least two materials for cushioning is needed in order to optimise both the compacting and long term endurance properties of the shoe. Different manufacturers design this key component of the shoe in different ways. For example, 'New Balance' encapsulate EVA within layers of polyurethane, while 'Nike' put air pockets in the polyurethane part. A recent innovation is to 'compression mould' the EVA such that it better resists shape change during the tens of thousands of load cycles imposed on the shoe. Another approach is to build a support bridge (Figure 60) into the shoe, giving in this case both longitudinal and lateral support where it is needed. Other types of polymer used in shoe construction include polyamide, polyester and certain types of elastomer.

Future trends in sports shoe design are likely to continue to emphasise the need to protect the wearer from injury to foot and limb. New, lightweight materials will make shoes lighter, more supportive and comfortable. They will also, hopefully, last better than today's products.

A tall order for manufacturers, perhaps; but it is after all a multi-billion dollar market out there.

Making a Racquet

It has recently been estimated that in the last five years the number of people playing tennis and squash has increased by almost 300%. The manufacture of tennis, squash and badminton racquets has thus become big business. Not so long ago, racquets were all made from wood; today it is almost impossible to find a wooden tennis, squash or badminton racquet in the store. First came aluminium frames, including the wide-frame (oversize) variety, to expand the 'sweet spot'. Before long, however, aluminium was forced to make way for glass or graphite fibre reinforced epoxy,

Applications

this providing better strength, lower weight and more durability. Today, other advanced fibre reinforcement such as kevlar and ceramics are being utilised in racquets with claims to dampen shock loading and hence reduce incidence of tennis elbow!

Up to a few years ago, the rules of squash prescribed that racquets simply had to be made of wood. However, squash is played in a rather confined area, and the danger to players by flying splinters from broken racquets forced a change in rules to allow stronger composite or metal alloy frames to be used instead. Indeed, this emphasis on designs to reduce injury is in itself good justification for the use of the new advanced materials.

It is an interesting fact that a tennis racquet requires between ten and eleven metres of string, depending a little on string tautness, of course. Not so long ago when ox-gut was used as string, as much as 80 metres of gut was needed to string a single racquet. Since an ox only has about 40 metres of gut, each racquet thus accounted for two oxen! Indeed, if it were not for the development of synthetic polymer fibre strings, there would simply not have been enough oxen to go round to supply the millions of racquets sold today.

A properly designed racquet dissipates a minimum of energy and allows the player to impart more velocity to the ball for a given effort. Thus racquet frames should be of high strength with high stiffness or elastic modulus and good shock dampening characteristics. Construction of such a racquet is illustrated in Figure 61. Most racquets are manufactured using a woven graphite prepreg (shaped like a woven cotton finger bandage). The graphite frame is pressurised internally using a rubber 'inner tube' and then cured in an oven. There can of course be several different possible materials combinations in frames, e.g. graphite/kevlar/epoxy; graphite/glass-fibre/epoxy; graphite/boron-fibre/epoxy; etc., depending on how much the customer is prepared to pay for the extra few percent of stiffness and dampening power in the racquet. Typically, the amount of fibres in a graphite/kevlar/epoxy composite racquet, for example, is about 285 grams in a proportion of 70% graphite, 30% kevlar. In very sophisticated racquets, even the torsional rigidity of the grip can be increased using a boron-fibre/epoxy composite.

The requirements for a squash racquet differ somewhat from

Tomorrow's Materials

61 High-tech racquet reinforced with Kevlar and glass fibres wound around a polymer foam core for stiffness, lightness and good shock dampening

tennis. Thus, lightness and stiffness in this case must be combined with considerable toughness to absorb energy from accidental smashes against the wall or floor. One sophisticated approach to achieve all this is to use an injection moulded foamed (cellular) nylon reinforced with chopped graphite fibres. Glass or kevlar/graphite/epoxy composite frames are also made in a so-called one-piece construction which is supposed to improve impact toughness, whilst retaining satisfactory stiffness.

As improved higher stiffness fibres become available, racquet manufacturers will no doubt continue to make the most of each additional gram of fibre that can be accommodated. Ceramic fibres such as silicon carbide and aluminium oxide (called fibre FP) are being investigated. However, potential customers, beware! Someone has calculated that the cost per kilogram of a high-tech racquet already compares well with that of a jumbo jet!

Applications

Easy Riders

Whether it be cycling to school or work, panting up mountain sides, participating in the Tour de France, flying record-breaking pedal-powered planes over the sea, or simply sweating on exercise bikes in the bathroom, just about everyone, everywhere, enjoys bikes.

Bicycles are probably the most energy efficient of all earthbound transport media. It is estimated that in terms of calorie consumption per litre of fuel, a cyclist achieves the amazing equivalent of about 500 km per litre, or a 50 times better economy performance than the average family car! It is hard to improve on this, of course.

Traditionally, bikes are made of steel. Steel is a relatively cheap material; it is also tough, weldable, easy to paint, and in the form of thin tubular constructions, steel bikes look nice, too. The steel used is alloyed with chromium, manganese and molybdenum, all of which raise the yield strength. In the latest technology, tubes of this material are drawn down to 0.33 mm thickness so that frame weight is minimised. Cold drawing of the tube dramatically increases its strength, too. On the other hand, lightweight materials such as composites based on graphite fibres and polymers, or light metals like aluminium, titanium and magnesium are attractive alternatives to steel in that they save weight. Comparing densities of these materials, aluminium alloys come out at around 3000 kg/m^3, the higher stiffness titanium alloys are slightly heavier at 5000 kg/m^3, while magnesium alloys weigh in least at 1500 kg/m^3. These values compare with the extremes of steel at 7000 kg/m^3 and carbon fibre/epoxy composites at 1500 kg/m^3. Of the metals listed, titanium and magnesium appear to give the stiffest competition to steel. Titanium alloys offer high strength-to-weight ratio and excellent corrosion resistance, and alloys based on the well-known aerospace material of Ti-2Al-2.5V are nowadays used in frames for mountain bikes which have to be tough enough to handle the roughest possible terrain. Magnesium on the other hand is the lightest of commercial metals (lithium is the lightest metal of all but is still only used as an alloying element, e.g. in aluminium alloys) with a density of only a third that of aluminium. Magnesium is attractive for use in

Selected Wonders of the World of Advanced Materials 6

The Bicycle Wheel

Considered by many to be 'man's greatest invention', the wheel continues to roll well abreast of the frontiers of modern science. In countries rich or poor, in Peking and Bombay, in Copenhagen and Singapore, the bicycle wheel is everywhere utilised by man. In many ways, bicycle wheels have become extensions of man himself, like the wearing of seven-league boots of old. In its ultimate form, the bicycle wheel is a highly developed composite material member of the ultimate of racing chariots: the competition bicycle.

A light, taut, well-balanced wheel is a must for the racing cyclist, since he wants as much energy as possible for propelling his machine. Racing wheels have used highly tensioned spokes which are radially straight (i.e. straight from hub to rim). More recently, however, ways of reducing aerodynamic drag of the wheel have tended to govern the development of bicycle wheels. First, elliptical and flat spokes were tried in conjunction with 'aero', V section rims. Since the mid to late 1980s, however, carbon reinforced composite materials have been used to manufacture the disc wheel. The philosophy behind the use of disc wheels is that the aerodynamic resistance due to spokes is removed. However the international rules of racing do not allow aerodynamic aids such as cowlings or covers to be used in competition. This was overcome by making the coverings structural. The disc wheel is basically a contoured sandwich structure consisting of two carbon/epoxy facings with a foam (polystyrene) or honeycomb (Nomex) core. The facings are bonded to flanges on an aluminium hub assembly and to a carbon/epoxy prefabricated rim. The core material is either bonded into place during this operation or injected at a later stage. The result, as seen in the figure, is no mean development from the wooden chariot wheels of ancient Rome!

The future, however, may see a return to spokes. The latest to be tried are three-spoked carbon/epoxy wheels, lighter than disc wheels but with lower drag than wheels with conventional (tensioned) spokes.

Applications

The ultimate in bicycles (and wheel) technology. Handlebar, frame and disk wheels are from carbon fibre reinforced epoxy; the forks are from aluminium. The rider is East German champion Jens Glucklich. Photo by H. A. Roth.

Tomorrow's Materials

bicycle frames because it die-casts well, machines easily and can be brazed without difficulty. One UK manufacturer has succeeded in refining a process for making magnesium bicycle frames using a special hot-chamber high pressure injection-moulding technique. By this approach, precise alignment can be achieved when producing thin-walled tubular sections, with good strength and torsional stiffness. These frames are currently produced for racing bikes, all terrain bikes and touring bikes.

Compared to steel, lightweight materials are all relatively expensive. Furthermore, there can be joining problems, particularly when using the fibre-reinforced materials, unless radically different design approaches are employed (see box figure, p.129). The high specific properties (property per unit density) of carbon/epoxy and carbon/kevlar/epoxy composites have, however, been utilised in the 1980s to produce lighter and stiffer frames. Initially filament wound carbon/epoxy tubes were bonded into aluminium lugs to produce a frame that still used metal (aluminium) forks. Improvements in winding and joining techniques have allowed the production of lugless frames which are manufactured completely by a filament winding technique. These frames have been used with great success by the East Germans in world championships. Carbon/kevlar/hybrid composites are also finding application in cycle frames and are now manufactured by a French company. The use of these composite materials is likely to proliferate in the 1990s because of their excellent strength to weight and stiffness to weight ratios.

As with other sporting goods, bikes need to be safe, reliable, look nice and hold a reasonable price. Expensive, lightweight bicycles will certainly become progressively more available in future, but exotic materials appear unlikely to dent the dominance of steel in everyday bikes for years to come.

Boards and Boats

Pottering about in boats or on boards is a pastime that few people in the world have not tried at one time or another. Whether it be a surfboard, a punt, an Indian kayak, a rowing dinghy or an Americas Cup yacht, the fascination is always there. New materials have played a large role in bringing boating to the masses,

Applications

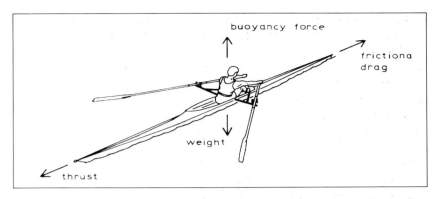

62 Forces acting on a boat. Modern materials can help reduce weight and minimise frictional drag

and while glass fibre reinforced epoxy boards and boats are definitely not as aesthetically attractive as reed rafts, canvas canoes or wooden dinghies, they are generally lighter, more portable, require less maintenance and are cheaper. At competition level, advanced (non-wooden) materials nowadays dominate board and boat construction for reasons of lightness, lower skin-friction, higher toughness and better safety.

Irrespective of the type of watercraft concerned, there are four basic forces to contend with when considering boat design. These are: weight, lift, thrust and drag. Weight is simply the gravitational force acting on the craft due to the weight of boat and passenger. In a lightweight single scull boat with a graphite fibre/epoxy resin shell, its weight can be as little as 10 kg! Referring back to Figure 62, lift is generated by the boat's buoyancy, i.e. the displacement of water by the craft's hull. On the other hand, when the boat moves, 'dynamic lift' occurs and this in turn reduces the buoyancy force. Thrust and drag (Figure 62), are forces that propel the boat along or resist the boat's motion due to friction and bow resistance. Obviously these resisting forces are mainly a function of the hull's design. However, certain materials can be incorporated in the hull which lower skin friction. Skin friction in this case is actually a fairly complex phenomenon and comprises of a smooth (laminar) component of flow and a turbulent (chaotic) one. Since turbulent flow increases drag, skin resistance can be

Tomorrow's Materials

reduced if the transition from smooth to turbulent flow is delayed as long as possible as the boat picks up speed. NASA have studied this phenomenon and found that waxing the hull's surface has little effect on skin friction. If, however, the hull's surface contains fine grooves running in the flow direction, drag can be reduced slightly. This effect was actually utilised in the 'Stars and Stripes', winner of the 1988 Americas Cup competition. By coating the aluminium hull with a fine-grooved polymer material, the yacht's speed was increased by 2%. Although this may not sound much, it proved to be pretty decisive in the finals, providing a four-boat-length advantage when measured over the whole course.

The key objective in boat design then is to minimise drag at the normal operating speed of the boat. The obvious approach is to have a shell that is lightweight, long and narrow. In the past, racing boat shells were beautifully crafted from cedar, spruce or mahogany and they were made lighter by constructing their hulls as thin as paper. These shells were prone to damage and many a careless finger penetrated the hull's shell during handling. The 1950's witnessed development of glass fibre/polymer hulls and by the late 1960's fibre-composites began challenging the dominance of wooden boats. Today, of course, wooden shells in rowboats are a rarity, indeed.

Surprising though it might seem, the materials science of 'simple' windsurfing boards is every bit as sophisticated as racing boats. As illustrated in Figure 63, the board's core consists of an extruded foam polystyrene filler enclosed in fibre-glass. Wound round the core are a number of spun graphite fibre strands embedded in a PVC resin matrix. This PVC/fibre composition is enclosed in tum by four layers of high-stiffness E-glass fibre weave. Finally, the whole is enclosed within a glass-fibre reinforced epoxy composite, with extra fibre strengthening of kevlar in parts likely to be exposed to extra wear and tear. Clearly, surfing boards of this complexity can only be produced by specialist companies. However, some enthusiasts like to build their own boards. Kits can be bought from which boards are constructed using extruded foam polystyrene (EPS) as a core material. This is then covered with glass-fibre (or Kevlar fibre) weave, built up and shaped with resin. Higher strength kevlar or graphite weaves can be used instead of glass fibre, but in that case there is

132

Applications

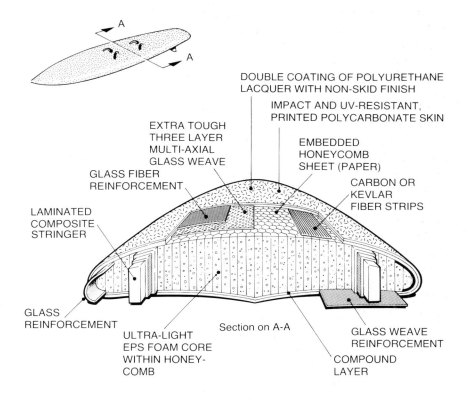

63 Cross section of a super-composite surfboard; surely the last word in advanced materials construction. (Courtesy of FANATIC).

a ten times cost penalty. More recently, simpler surfboard constructions have been made by using a mouldable polyethylene copolymer. This foamed polymer reduces weight and avoids time-consuming hand shaping operations. It should also be cheaper.

Surfboards are not the only watercraft that need to be tough. When steering kayaks or canoes through rock-strewn streams it might well feel safer to be in a toughened polymer foam and fibre composite hull than a canvas craft as in traditional Indian style.

Tomorrow's Materials

Indeed, modern canoe hulls typically consist of sandwiches of foamed polythene encased in cross-linked polythene sheet, with ethafoam as protection padding. A moulded polythene hull can easily be made smooth and shaped for good water dynamics. Furthermore, the whole is a fraction of the weight of the wood and canvas models of yesteryear, and hence more portable.

The future is likely to see even more use of the very advanced graphite/aramid/kevlar fibre-reinforced composite materials in hulls, paddles and oars. A somewhat revolutionary new light-weight polymer with higher impact resistance than aramid, currently being tried in boat hulls, is called 'Spectra'. This is a gel-spun, extended chain polyethylene, produced as a weave, and has a higher specific strength than any other organic fibre available. Hulls made from this material will be lighter, tougher and more wear-resistance than other composites. In spite of all this, however, the traditions of boating being what they are, there is a growing section of enthusiasts who actually prefer crafted wooden hulls to the oven-baked jobs, and are quite prepared to sacrifice time and energy on the scraping and varnishing of planks or the sewing of canvas.

Skis go Cellular

There is a wonderful sense of freedom and fresh air when negotiating a mountain slope or following a cross-country trail on skis. Twenty years ago, skiers used fairly heavy, monolithic constructions made of wood. Today's equipment is based on some of the most advanced design and materials science principles known.

In order to understand ski construction it is useful to consider some of the basics involved in their design. As illustrated in Figure 64, the forces acting on skis are gravitational (weight of the skier), lift or buoyancy due to the shape of the skis and the way they 'ride' the terrain, and the two frictional forces of air resistance and friction between snow and skis. The aerodynamic force depends, of course, on the skier's bulk, his technique and style. The other forces are all dependent on the form and construction of the ski, as affected by the skier's weight and speed. The best glide conditions are met when the skier's weight is distributed equally over the contact running surface of the ski. This pressure

Applications

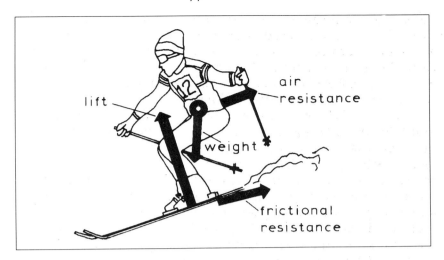

64 Forces acting on a skier

will, in turn, be affected by such factors as speed and turning radius (hence skill) of the skier. In other words, the thickness and stiffness of a ski should ideally sensitively reflect the weight, strength and ability of the skier. Under- or overloading of the ski may cause 'chatter' and poor edge control during turns. In cross-country skiing, skis must be as light as possible, yet accommodate undulations as well as large variations in snow temperature and firmness in the track. Both downhill and cross-country skis must also be tough enough to absorb shock loading, and overcome vibration due to gliding dynamics. To summarise, then, skis need to possess both longitudinal and torsional rigidity to ensure good allround weight/pressure distribution, edge holding ability, and flexibility over a bumpy terrain.

The characteristic profile of a ski is illustrated in Figure 65. The degree of camber and stiffness of the ski accommodate, as far as possible, the various loading conditions discussed. Thus, the ski's flexibility is a function of the skier's weight and the various section thicknesses, from the ski's centre section (rigid) to its tip/tail (flexible). In a monolithic, homogeneous ski construction, it is obviously difficult to optimise all of these factors. In spite of this, the early monolithic skis, constructed and shaped from high strength woods such as ash or hickory, performed well. As a

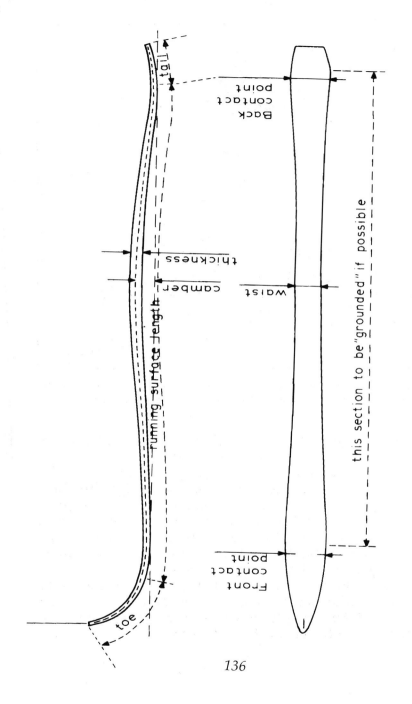

65 Profile of a downhill ski constructed to accomodate the weight of the skier in motion. By courtesy of Dr. Hugh Casey, Los Alamos National Laboratories

Applications

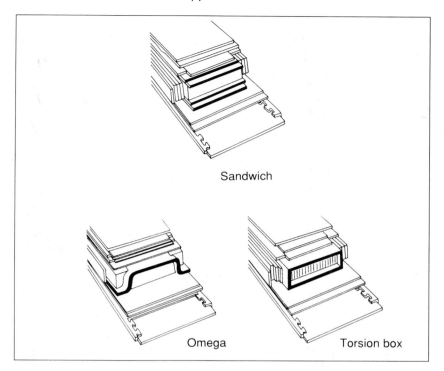

66 Different types of construction of downhill skis. In practice, combinations of all three can be used. By courtesy of Dr. Hugh Casey, Los Alamos National Laboratories

cellular material, wood possesses good properties of flexibility and toughness, and is easy to treat by waxing or varnishing. Indeed, wood is still a popular core material for skis.

Examples of some modern composite ski designs are illustrated in Figure 66. The 'sandwich' construction allows good control over longitudinal flex, the 'torsion box' increases torsional rigidity, while the 'omega' design tends to combine both as a function of 'omega' size. In reality, most skis today utilise combinations of all three of these design concepts.

It is seen in Figure 67 that a number of different materials make up the cross-section of skis. Thus the core of the ski is a cellular material based, for example, on wood or an aluminium honeycomb or even a polyurethane foam. In more advanced downhill

Tomorrow's Materials

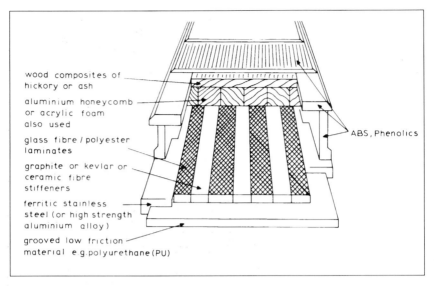

67 Example of the different materials used in a downhill ski

skis, acrylic foams (very light, good dampener), and more recently kevlar or graphite fibre/epoxy honeycombs are also used. The sandwich construction surrounding the lightweight cellular core of the ski thus provides stiffness and flexibility to the ski. As illustrated in Figure 68, by using different combinations of materials, e.g. from high strength steel to advanced fibre composites, properties such as stiffness, strength, toughness and elastic response of the skis can all be controlled in a fairly precise way.

Materials that may improve and optimise the skis' characteristics even further will undoubtedly appear in future. Tomorrow's skis will no doubt be lighter (using new advanced polymer foam cores), possess better damping characteristics (by using ceramic fibre reinforcement), will better accommodate both flexural and torsional stresses (using improved and more sophisticated design techniques), and adapt better to different gliding conditions (by the use of superior coating techniques). As with all the other applications of materials in sports equipment, ski manufacturers will continue to optimise products just as far as the customer's ability (and pocket) allows.

Applications

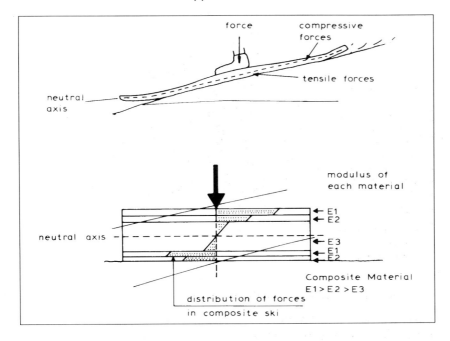

68 Schematic distribution of forces in a composite ski

Future Trends

In this section, we have briefly considered only a few areas of sport and sports equipment. There are many other areas equally active in the application of advanced materials. For example, interesting progress is being made in the upgrading of wood to advanced fibre composites, in racing oars and double-bladed paddles for canoes. Ski poles, ice hockey clubs, fishing rods, table tennis bats and windsurfing masts are all in the process of redesign, again mainly based on polymer-composites. Even the 'untouchables', cricket and baseball bats, are being eyed by materials scientists as ready for 'up-grading'. The dominance of willow or hickory wood is all very well; but think how built-in springs of carbon/kevlar/ceramic composites could improve power, dampen shock loading, and perhaps double up on the sixes and home-runs!

Tomorrow's Materials

Further reading

Popular accounts of the applications of advanced materials are to be found in popular scientific and technical magazines such as *New Scientist* and *Scientific American*, as well as more specialist journals such as *Advanced Materials and Processes* (published by ASM International). Particularly recommended is the October 1986 issue of *Scientific American*, which is devoted entirely to new materials.

PART III
TOWARDS TOMORROW'S WORLD

The Earth's natural resources are generally defined in dictionaries as "forest and minerals that can be drawn on". If such resources were compared with savings in the bank man has, during the past century or so, indulged in one gigantic binge. What is the damage? Can anything be done? What role do tomorrow's materials play in all this?

Tomorrow's Materials

Planet Earth: Ecosphere or Dustbin?

In the last decade or so man has shaken himself as if from a drunken stupor, and seen with horror the depleted state of his resources, tasted the pollution of his surrounds, and witnessed the diminishing (or disappearing) wildlife around him. The environmental issue has suddenly become a hot subject for the media and politicians' speeches; yet most environmentalists fear that too little is being done, too late.

Consider briefly the problem. Prior to man, Planet Earth was green and beautiful and full of life because of a delicately contrived ecological system (an ecosphere) in which a dynamical balance existed between the cycles of life and death, and the surrounds of soil, sea and sky. Modern man, with his superior attitude and knack of claiming (and defending) ownership of pieces of the planet, soon developed uses and commodities out of the plants, soil, stones, and animals within his domains.

Take the forest, for example. It is estimated that already in prehistoric and medieval times, destruction of forest by man occurred at an average annual rate of half a million hectares per year (give or take a few tens of thousands of trees). Most of this was taken from the temperate forests of Europe and North America and used as building materials for houses and boats, for firewood, or it was simply torched to clear land for crop growth. In more recent times, however, the evergrowing demand of the "advanced" countries has caused man to turn to the rainforests of South America, South East Asia and Africa. Here, deforestation is currently occurring at the dizzy rate of 20 million hectares a year. As of now, roughly 20% of all the planet's plant and tree life has been demolished. This is roughly the equivalent of all of the Earth's cultivated cropland, amounting to same 800 million hectares.

As an isolated example of this rampage, the following extract is from a recent issue of *New Scientist* (26 November, 1988):

> "At the beginning of this month a cargo ship carrying 11,000 cubic metres of logs began unloading at the Japanese port of Kitakyushu. The logs were from the

Towards Tomorrow's World

Amazon rainforest, a new source of material for Japan's timber industry, the world's largest consumer of tropical hardwoods. A further 27,500 cubic metres will arrive in December; next year, the shipments should be running at the rate of 40,000 cubic metres per month. About one-fifth of the timber will end up as wooden boards for shaping concrete on building sites, and will be discarded after a few uses."

Not all deforestation is due to logging. Huge areas of rainforest are cleared by hacking and burning for croplands and to accommodate the recovery of iron ore, and even gold. Rainforest soil is actually very thin and unsuitable as cropland, so only the poor tend to farm there. Worst of all, the clearing of rainforests like this locally upsets the ecological cycle so that heavy rain quickly brings about soil erosion and yields the land barren.

Deforestation of the magnitude now going on worldwide contributes 30% of all the carbon dioxide that is emitted to the atmosphere; the other 70% comes from the burning of fossil fuels such as coal, oil and natural gas. Add to this emissions of nitrous oxide and carbon monoxide from combustion engines, coal-fired power stations and even fertilisers, and we "achieve" a staggering (estimated) 5,000,000,000 tonnes of dirty gases which spew into the air annually. The net result of this is actually quite alarming: it creates a "greenhouse" effect on Earth, blocking the ecological cycle and causing the atmosphere to heat up (see Figure 69). If present trends continue, as indicated in the figure, a temperature increase (measured aver 50 years) of over 2°C will have occurred this century. This may not sound much, but experts claim that such an upset to the environment could threaten the existence of a great many of our "fellow" plant and animal species.

The main threats to achieving a better environment are: the continued growth of man (10 billion by the middle of the next century?), the waste products that he produces (sewage and industrial), his greed for consumer goods and his thirst for energy. In the following pages we need also to consider the role of tomorrow's materials in the light of all this; are they environment-friendly or do they merely help satisfy man's crave for consumerism? And what, if anything, can be done?

Selected wonders of the world of advanced (natural) materials 7

The rainforests

The rainforests have been described as the Earth's richest treasure trove of living creatures. For example, a single hectare of the Tambopata reserve in Peru, illustrated here, probably contains over 800 species of woody plants. Three-quarters of all species that creep or swim or fly, including 90 per cent of the Earth's insects, live there. There are 570 species of birds, 1200 types of butterfly, 600 species of leaf beetle, 127 reptile types and some 90 different mammals, to mention just some of the species found there.

One-third of all the world's rain forest is in Brazil; Madagascar accounts for a quarter of Africa's plant species; Colombia has the world's richest diversity of birds and Indonesia its highest mammal count. Zaire has Africa's largest remaining tract of rain forest.

If left undisturbed, the rainforests form an "ecosystem" in which plants and creatures live complex inter-dependent lives. In effect the system equilibrises the delicate balance between the life cycles of plants and living creatures, and the environment. Rainfall and sunshine are distributed in delicately divided domains of the forest. Waste products and rotting trees return to the soil and replenish it. The whole, the *ecosystem*, is perhaps the greatest natural wonder of the world we live in. Today, its function is seriously threatened by rampant deforestation projects.

Towards Tomorrow's World

The Tambopata reserve in Peru, considered to be one of the world's richest treasure troves of life. (Photograph by Anna Culwick (TReeS).

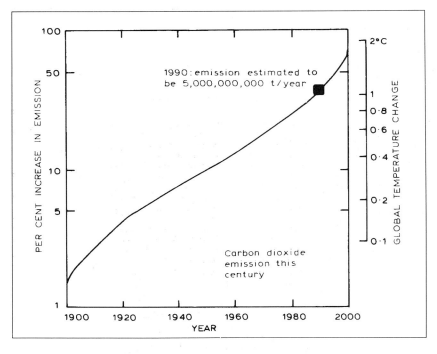

69 Increases in carbon dioxide emission into the atmosphere with corresponding effect on global warming. After: The Ecology of Natural Resources" (I G Simmons; publ. by Edward Arnold, 1981)

The Strategic Elements

Concentrations (by weight per cent) of same of the elements that occur in the Earth's crust, its seas and atmosphere are shown in Figure 70.

It is interesting to note that 99 per cent of mineable minerals (from the crust) are made up of only eight elements. Of these, the important metallurgical elements, aluminium and iron, are clearly in very good supply. Of the remaining elements listed in the figure (bringing the total to 99.89 per cent) the metallurgically important are titanium, manganese and carbon and these appear also to be relatively abundant. However, metallurgically important elements within the remaining 0.11 per cent of the crust

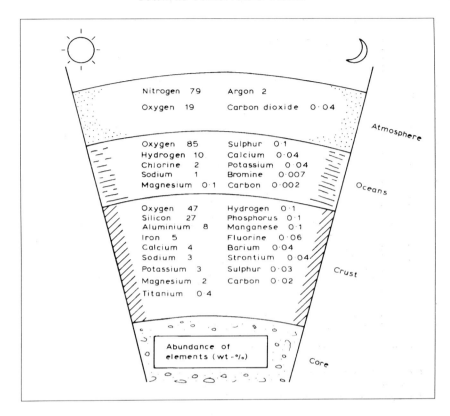

70 Comparison between relative abundances of recoverable elements in the Earth's crust, oceans and atmosphere. The amounts in each region are estimated to be very approx. 10^{20} tonnes. (From various sources)

include: vanadium, cobalt, nickel, copper, zinc, niobium, molybdenum, silver, zirconium, cadmium, tin, tungsten, platinum and gold. Indeed, as we shall see, many of these elements are in short supply. It is worth noting from Figure 70 that most of the constituent elements of the new advanced ceramics such as aluminium oxide, silicon nitride, silicon carbide, (SiAlON), etc., are amongst the most abundant on Earth.

We can define the strategic value of an element by a number of factors such as: its abundance and application; how much of it has to be imported; from which country it is purchased (friendly or otherwise). Figure 71 campares same strategically important ele-

Tomorrow's Materials

ments in terms of their source countries, for the cases of the USA and the USSR. Apart from the fact that the USA is seen to be much more dependent upon imported materials than the USSR, certain strategic metals such as manganese, chromium and platinum have to be imported to the USA from South Africa and the USSR. Likewise, the Soviet Union is quite independent of the USA and South Africa. It is an interesting fact that five countries in the world (USSR, South Africa, USA, Canada and Australia) actually produce over 90 per cent of the materials listed in Figure 71 (p.150).

The strong dependence on imported materials by both the European community and Japan is shown in Figure 72 (p.152). Note the heavy dependence of these areas on such important strategic metals as copper, manganese, chromium and nickel. Furthermore, even if the EEC and Japan are themselves producers of iron and steel, most of the iron ore used by them has to be imported. On the other hand, countries such as the USA and Japan, and the EEC countries are all producers of high-tech commodities where the value of the materials is greatly enhanced. The value of exported materials is low, by comparison.

The relative scarcity of certain metals is illustrated in Table 10. In the final column of this table, the ratio of expected resources relative to their likely demand shows that the supply of at least four important metals (copper, lead, molybdenum, tungsten) will barely keep abreast of their demand. In the case of silver, used extensively for example in photographic emulsions, only 40 per cent of demand is predicted to be met by the turn of the century. Other industries that obviously need to be concerned over resource supply include cable manufacturers (substitution of copper by aluminium is in progress), and producers of cemented tungsten carbide tools and drills (ceramics and ceramic coatings may be substituted here, perhaps).

Energy and population

Three of the most fundamental sources of harm to the environment are the sheer size of the population on Earth and its thirst for

Towards Tomorrow's World

Table 10   Demand between 1974 and the year 2000, and reserves of certain common minerals

Minerals	Units in metric tons	World demand 1974–2000	World reserves in 1974	Reserves in relation to cumulative demand 1974–2000
Aluminium	Million	873	3483	4,0
Cobolt	"	1,16	2,45	2,1
Copper	"	320	408	1,3
Chromium	"	92	523	5,7
Gold	Thousand	32,8	41,1	1,3
Iron ore	Million	20 000	91 000	4,5
Lead	"	125	150	1,2
Manganese	"	370	1826	4,9
Molybdenum	"	4,08	5,90	1,4
Nickel	"	26	54	2,1
Phosphate	"	6175	16 065	2,6
Silver	Thousand	420	187	0,4
Sulphur	Million	2036	2040	1,0
Tin	"	7,65	10,32	1,3
Tungsten	"	1,48	1,78	1,2
Zinc	"	217	236	1,1

Source: US Bureau of Mines/United States Department of the Interior, "Mineral Facts and Problems", 1975 Edition.

food and energy. It is predicted that during the next century the world's population will achieve 10 billion people. To supply food, water and firewood for what amounts to over twice the number of people alive today will necessitate changes in attitudes to conservation of energy and resources, not to mention supply of food and sewage problems. Even fairly conservative estimates of the life cycle of crude oil production in the world predict that 80 per cent of oil will be exhausted by about the year 2030.

The main fossil fuels used today are coal, oil and natural gas. The projected lifetimes, based on a modest 5 per cent per year growth rate, are shown in Table 11. It is seen that whilst there is a good supply of coal, oil and gas will not survive the next century. An example of recent predictions of Britain's energy uses up to the year 2020 are illustrated in Table 12. As shown here, nuclear energy will drop by 14 per cent, but coal and natural gas use is

149

Tomorrow's Materials

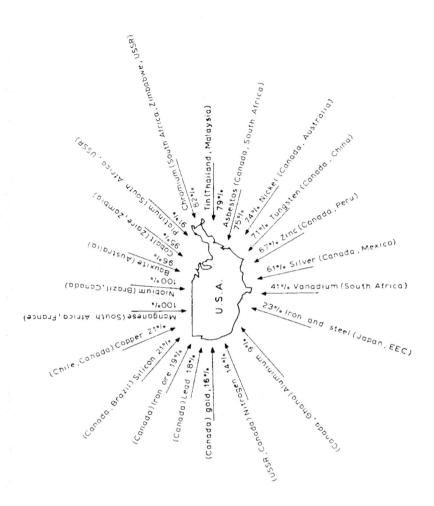

Towards Tomorrow's World

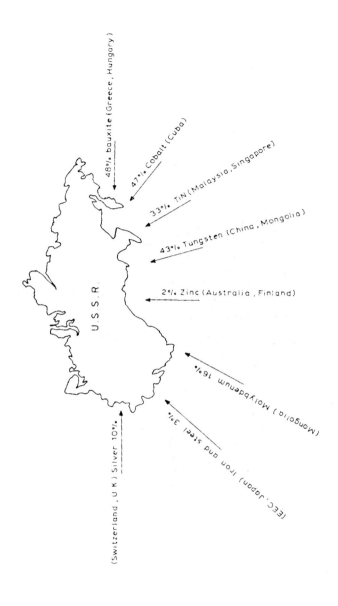

71 Comparison of main imported materials to the USA and USSR, and their main suppliers. Note the much heavier dependence of the USA on imports. (From various sources)

151

Tomorrow's Materials

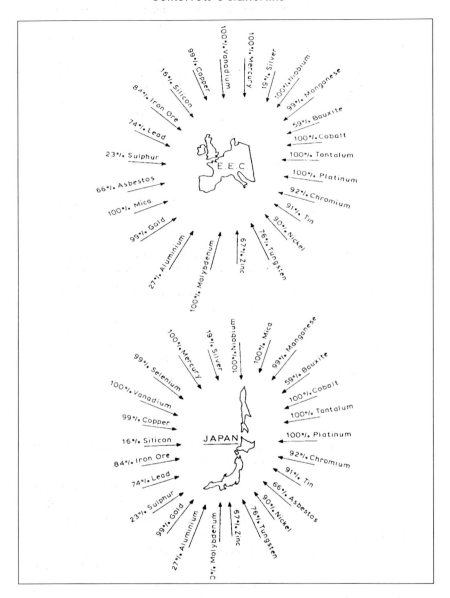

72 Comparison of main imported materials to the EEC and Japan. (From various sources)

Table 11 Conventional fossil fuels: reserves, resources and supply

	1 Proved reserves (EJ)	2 Remaining URR (EJ)	3 Recent annual consumption (EJ)	4 Reserve lifetime (col 1 ÷ col 3)	5 Resource lifetime at growth rate increase of 5% p.a. (Yr)
World coal	124,000	220,000	90	1,380	96
World petroleum	3,700	10,900	100 (136 in 1978)	37	37
World natural gas	2,100	10,600	40	53	53

Source: I. G. Simmons, *The Ecology of Natural Resources*, Edward Arnold, 1986.

Table 12 Britain's total energy use in millions of tonnes of coal equivalent

	1985	1990	2005	2020	% up on 1985
Coal	105	115	131	174	66
Oil	114	118	145	165	45
Nuclear	22	25	32	19	− 14
Hydroelectric	2	7	7	7	
Natural gas	77	78	120	145	88
Renewable energy sources	0	1	4	8	
Total	320	344	438	517	
Carbon dioxide emissions resulting (millions of tonnes)	601	640	819	1037	

Source: *The Independent*, London, November 1989.

expected to increase substantially. This result may be promising for the anti-nuclear brigade but is decidedly bad news for prospects of reducing carbon dioxide emissions, as emphasised horrifyingly in the bottom column of Table 12.

A world that has grown used to the freedom of cheap travel, the comfort of warm houses, the convenience of electricity and gas supply will not willingly pay more for its power supply or accept that it should be used more frugally. On the basis of current technology, almost all cures to energy problems, e.g. replacing nuclear energy by fossil fuels, involve aggravating the environ-

153

Tomorrow's Materials

Table 13 Energy consumption of different passenger transport methods

Mode	Net energy consumption at average occupancy: kWh/passenger kilometre
Bicycle (tandem)	0.01
2-stroke moped	0.20
Small private car in town	0.50
Large private car in town	1.06
Bus (urban)	0.25
Subway (electric)	0.11
Diesel train ($<$ 100 km trip)	0.29
Electric train (long distance)	0.09

From various sources.

ment even further. Only energy conservation and population control would appear to offer an answer. By conservation, is usually meant the use of a lower quantity of energy per capita but without drastic changes in lifestyle. Materials solutions of this type could include the use of cellular insulation in houses, double or triple glazing in windows, more efficient lagging of pipes and roof spaces, the use of lightweight cellular structures in trains and vehicles, more efficient (ceramic) engines, etc. Examples of the energy consumption of different passenger transport vehicles are given in Table 13, well illustrating same extremes in our ways of expending precious energy. Yet another approach towards materials conservation is through recycling.

Recycling: the Materials Merry-go-Round

Relatively few materials are actually destroyed in the course of their use, so that given the appropriate cost structure, the recycling of materials is an attractive and viable way of saving resources and energy. Metals, wood and paper are obvious examples, but the approach can be applied to many other materials too, e.g: lubricating oil, water, sewage, textiles, plastics etc.

Towards Tomorrow's World

Obviously, the success of a recycling process depends on how much of an item can be recycled. For example, an obsolete machine may have high scrap value, but iron rods embedded in concrete would be hard to recover. An example of the sources and amount of recycling of metals in the USA (as of 1966) is shown in Table 14. It is seen that recovery of iron and steel scrap was actually fairly high (*ca.* 50 per cent); this can derive from several sources, as indicated in Figure 73.

One of the problems of recycling metals is that the manufactured items, e.g. automobiles, contain fairly large amounts of other (unwanted) materials. A typical materials merry-go-round is illustrated in Figure 74, showing how the contaminant level can grow during the cycle. Obviously, methods of reliable analysis and appropriate blending have to be developed if the "renewed" metal is not be badly contaminated. A particularly serious contaminant in the case of auto-recycling, for example, is copper. An automobile contains about 1 wt per cent copper, and this level has to be reduced by about 90 per cent using appropriate "shredding" procedures if the scrap is to be acceptable by the steel manufacturer.

Steel is, of course, not the only metal to be recycled. ALCAN plan to build a special plant for recycling aluminium cans soon, claiming that recycling needs only 5 per cent of the energy required to produce aluminium from bauxite. Du Pont have also announced plans to recycle plastic fram municipal wastes, claiming that: "plastics must be the most recycleable of all materials". Welcome news indeed!

The recycling of paper products is now expected to be approaching 30 per cent for the case of newspaper. And well it might! Surely one of the worst examples in this respect is the Sunday edition of the *New York Times*; Larry Hirsch of the Norfolk Virginia Pilot has written:

"I think I shall never see
A *Times* much thinner than a tree.
A weekday's *Times*, although also big,
Compared to Sunday's is a twig."

This poem, part of a brochure aimed at *promoting* the *New York Times*, could well serve as a sad epitaph to Earth's fallen forests.

155

Tomorrow's Materials

Table 14   Scrap metal in the United States as of 1966

Metal	Approximate annual recovery from scrap (1 ton = 1.016 tonne (metric) 1 oz = 28.349 g)	Remarks
Iron	70–85 million tons	In the iron cycle from mine to product to recovery, the loss of iron is 16–36 per cent. About half the feed for steel furnaces is scrap.
Copper	1 million tons	Secondary copper production from old and new scrap ranges from 900,000 to 1,000,000 tons per year, about half of which is old scrap. Old scrap reserve is estimated at 35 million tons in cartridge cases, pipe, auto radiators, bearings, valves, screening, lithographers' plates, etc.
Lead	0.5 million tons	Estimated reserve is 4 million tons of lead in batteries, cable coverings, railway car bearings, pipe, sheet lead, type metal.
Zinc	0.25–0.40 million tons	Zinc recovered from zinc, copper, aluminium and magnesium-based alloys.
Tin	20,000–25,000 tons	Tin recovered from tin plate and tin-based alloys, 20 per cent; tin recovered from copper and lead-based alloys, 80 per cent.
Aluminium	0.3 million tons	This amount is growing rapidly. Nowadays 40–50% of aluminium is recycled, e.g. from cans.
Precious metals	gold = 1 million ounces silver = 30 million ounces	Previous metals including platinum are recovered from jewellery, watch cases, optical frames, photo labs, chemical plants. Because of the high value, recovery is high.
Mercury	10,520 flasks	Recovery is high. Nearly all mercury in mercury cells, boiler instruments and electrical apparatus is recovered when items are scrapped. Other sources are dental amalgams, battery scrap, oxide and acetate sludges.

Source: P. Flawn, *Mineral Resources*, 1966.

Towards Tomorrow's World

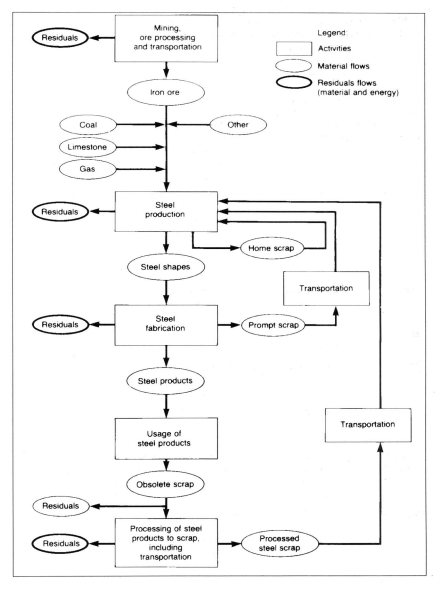

73 Schematic of steel production and fabrication, including recycling. (From "Automotive Scrap Recycling", by J W Sawyer, Resources for the Future, Inc., 1974)

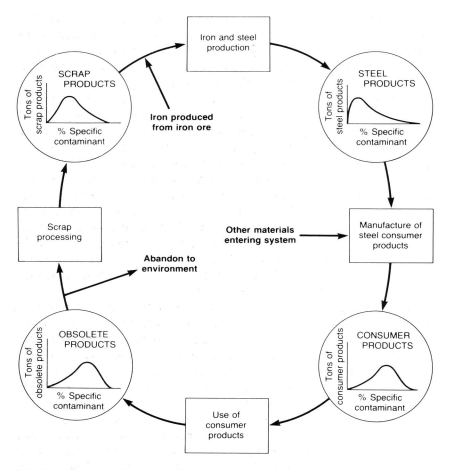

74 Schematic illustrating some of the ways materials can accumulate contaminants during recycling. (From "Automotive Scrap Recycling", by J W Sawyer, Resources for the Future, Inc., 1974)

Biodegradeable polymers: Drops in a plastic bucket

The public's aversion to streets, parklands and beaches strewn with tatty, disused plastic throwaways has at least prompted

Towards Tomorrow's World

some companies to look for ways of making self-destructive plastic products. Present production is very low compared to the huge amount of plastics made (mere 'drops in a plastic bucket'), but hopefully, interest in biodegradable plastics will increase now that ecology-conscious people are demanding that shops switch from plastic to paper bags. The Italian government is playing a leading role in this respect by stating that non-biodegradable packaging material, including shopping bags, will be phased out over the coming decade. Recently, Sweden has also outlawed packaging material made from non-biodegradable PVC.

The problem of developing biodegradable plastics is a challenging one to materials scientists. There have been several approaches to the problem, although few successes have been forthcoming. Britain's ICI has produced a biodegradable polymer called "PHB". It is produced by micro-organisms and breaks down by the action of bacteria. Unfortunately, it is so expensive to produce that it can only be used in very special applications, such as in surgical sutures (dissolvable stitches).

A more promising (cheaper) biodegradable polymer has been produced by a Canadian campany and is called "Ecolyte". Polystyrene cups are made by this company, such that when they are exposed for about 60 days to sunlight, they break down into dust particles. Bacteria then turn this dust into water and carbon dioxide. Currently, about 20 tonnes of Ecolyte are produced monthly and the material appears quite unaffected by long storage times, provided it is kept away from ultraviolet rays.

An Italian campany has recently produced a very promising biodegradable polymer which may be suitable for the large disposable carrier bag market. This polymer is based on a material consisting of webs of a short-molecule oil-derived polymer, heavily impregnated with maize starch. In time, these short-molecule polymer chains dissolve in water, while microbes degrade the maize filler to water and carbon dioxide. This solution is actually an elegant piece of materials science in that the oil-derived and starch polymer chains actually blend as a "polymer alloy", thus forming a material of new and attractive strength properties, appropriate for shopping bags. Indeed, this "alloying" is the secret to the high amount of starch filler that can be absorbed into the plastic, rendering it both mechanically strong *and* biodegradable. Current cost: approximately twice that of a conventional plastic bag.

The crunch: congestion or consciousness?

In a world such as ours, the creation of waste products and pollution is inevitable: food becames sewage, automobiles turn into junk piles, quarries and mines become derelict land, combustion engines and industrial chimneys spew smog and produce acid rain. A distressing graphical example of the inevitable industrial end-products of our most advanced industrial society is presented in Figure 75.

There is no such thing as a global garbage bin. If waste cannot be dug into the land, it must be tipped into the sea or burned. Can governments get polluters to pay for the dirt they create? This is like asking spoilt children to change their ways and give up their goodies for the sake of kids in poorer neighbourhoods, or even those yet unborn. It might even be suggested that most industrial countries are controlled by people with the most to gain from expanding consumer production, regardless of its consequences. And so the global garbage grows.

There has to be an alternative to this sorry state of affairs. For some time now, environmental researchers have been discussing the concept of an "ecosystem", in which dynamic equilibrium is maintained by holding populations relatively stable, and in which all materials are conserved by recycling them or returning them to their source. Obviously, to really achieve this state of being, the whole social consciousness of society also needs to change. Such fundamental changes begin with us, ourselves. What can we do?

A recurring theme in this book is the thesis that tomorrow's materials are more environment-friendly than today's. There are many negative accounts in the ledger, but if we try to be optimistic it should be possible in future to be more selective in the use of available resources. For example, the advanced ceramics derive

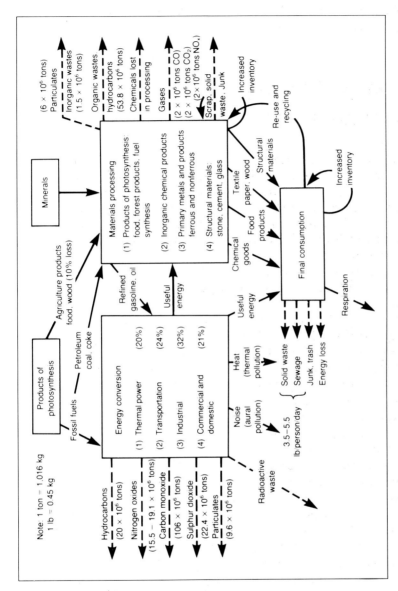

75 Schematic of material flow through the various social and industrial processes in the USA, illustrating how inputs of energy and materials became converted to waste. (After The Ecology of Natural Resources by I G Simmons, Edward Arnold, 1981).

Tomorrow's Materials

from the most abundant of elements; steels can easily be designed to avoid the use of scarce constituents; concrete beams could be cheaper and less energy-consuming than producing steel. Savings in fuel can be achieved by more utilisation of light cellular materials in transport media (a leaf out of Nature's book). More efficient (ceramic) heat engines can be developed from powder materials, and powder parts can be produced without wastage in achieving final shape. Sophisticated surface treatments can enable cheaper, less energy-consuming structural materials like mild steel to be more utilised in future. More efficient optical fibre materials can help avoid use of valuable copper in telephone cables. Evaporated polycrystalline silicon can be used for solar energy storage cells. New electronic materials can improve the world's communication systems, and help reduce paper in information technology; they may even help save electricity in the form of under-city superconductor power lines.

As one of our most sophisticated cellular materials, wood should be used sparingly and given the respect it deserves. Plastic products that are biodegradable should become the rule rather than the exception. Machines and parts should be designed such that when scrapped, the shredding and dividing of parts for recycling purposes is readily facilitated.

All this needs, of course, radically new concepts in materials science and design engineering; it also provides a tremendous challenge to our techno-eco-citizens of tomorrow's world! Instead of combatting the environment (like we do today), they can be environment creators; as in the spirit of Goethe's Nature:

"She creates forms ever new.
What is, never was before,
And what was, never returns.
All things are new, Yet it is always the old."

Towards Tomorrow's World

Further Reading

There are whole "Green" sections in most bookshops and libraries.
these days. However, for a good, comprehensive coverage it is still
hard to beat The Ecology of Natural Resources (2nd edition), by
I. G. Simmons, published in paperback by Edward Arnold, 1981.

Tomorrow's Materials

Glossary

alloy, alloying An alloy is formed by the mutual solution within a single phase or crystal structure of two or more elements. Even in alloys where several phases coexist, all the phases present are likely to be *alloyed*, i.e. contain two or more constituent elements.

amorphous Material with a random or non-symmetrical atomic arrangement. In metals and ceramics it means a *non-crystalline* state; in polymers it means a random arrangement of the atomic chains of the polymer.

anisotropic This refers to asymmetrical properties or shape of a crystal lattice or cellular structure. Thus the individual cells of wood are asymmetrical in shape, and have different stiffnesses when loaded in different directions.

austenite, austenitic The high-temperature phase, or (face centred cubic) crystal structure of iron and steel. When steel is hardened, it must first be annealed in the austenitic condition where its carbon content is easily dissolved.

binary alloy An alloy with two constituents. For example, steel is a binary alloy of iron and carbon.

biodegradable This refers to synthetic polymeric materials or compounds which are capable of disintegrating to some natural form or substance.

bioglass This is a fairly complex ceramic, based on silica glass, developed for its strength, compatibility, stability and function in the form of implant devices in the human body.

biomaterial Synthetic material used for implants in the human (and animal) body.

blend A polymer material containing two or more constituents within a given phase. Blends are also referred to as *polymer alloys*.

brass This is a single phase alloy of approx. 50 % copper and 50 % zinc. Brass is important commercially because of its corrosion resistance and its pleasant glossy appearance.

carbon steel All steels contain carbon (the definition of steel is that it is an alloy of iron and carbon), but carbon steel is usually of a ferritic form at ambient temperature.

cell In materials science, an open or closed symmetrical unit of a

164

Glossary

cellular material such as the single cell of a bee's honeycomb, or of wood.

cermet A composite of ceramic and metal, such as silicon carbide fibres in aluminium.

composite A material consisting of at least two separate phases or components such that the whole has properties superior to the individual components. Examples are reinforced concrete and glass-fibrereinforced polymers.

co-polymer A composite of two or more polymer types.

corrosion This is a phenomenon in which certain environments, e.g. salt water, attack and (through an electrochemical reaction) break down the thin oxide films that normally protect metals in air. Corrosion resistant metals such as stainless steel contain sufficient chromium that a hard tough film of Cr_2O_3 quickly forms at the surface, thereby protecting the metal under normal wear and tear. Corrosion is, however, still one of the largest of all problem areas in metallurgy.

creep A form of *plastic deformation*, usually occurring at quite a high temperature and over a long period of time.

crystalline On solidifying, the atoms of many materials form up in an almost perfect symmetrical lattice structure. Almost all metals are crystalline; other classical examples of crystals are diamond and ice.

deep drawing A manufacturing technique in which metal is shaped by drawing in a special die to its final form.

dislocation A linear discontinuity in the perfection of a crystal. A sufficiently large external force acting on the crystal may cause the dislocation, or groups of dislocations, to move, and this is the basis of *plastic deformation*.

dopant, doping A very small amount of an alloying addition which, when dissolved by a material, may modify its properties. An example is the doping of silicon by phosphorus or boron to make it semiconducting.

ecosystem This is an ecologically ideal environment in which plants and creatures live together in an enclosed system which equilibrises the delicate and complex balance between the cycles of life and death, such that the system is continually nourished and maintained. An example of an ecosystem is rainforest.

Tomorrow's Materials

equilibrium The stable state of a phase or structure. These stable states are illustrated for most binary and many ternary alloys in the form of equilibrium diagrams.

extrude A metalworking process in which metal is forced through a die or orifice.

fatigue A form of failure or fracture in a material produced by cyclic loading which can occur at loads well below the normal (static) strength of the material.

ferrite The low-temperature (body centred cubic) phase of iron. In most steels it is the stable phase at ambient temperature.

fine-grained microstructure Most crystalline materials are *polycrystalline,* comprising of many small crystals separated by crystal or grain boundaries. A fine-grained microstructure is a polycrystalline material with a fine grain size.

glass transition temperature A term used in polymer science for the temperature at which a change in stiffness or modulus of the material occurs. These polymers should not normally be used above the transition temperature.

grain An individual crystal within *polycrystalline* (multi-grain) material, in which adjacent grains have substantially different orientations.

grain boundary Each grain in a polycrystalline material is separated by a grain boundary about two atoms wide.

hardness An intrinsic property of a material, usually measured by an indentor of a specific shape and loading.

heat affected zone When metal is fusion welded, the areas near the hot part of the weld metal will be heated. This heating cycle can cause substantial changes in microstructure (e.g. grain size) and properties.

hot isostatic pressing A process for densifying ceramic or metallic powders under conditions of high temperature and isostatic pressure.

impact toughness *Toughness* refers to a material's resistance to crack growth. Impact toughness is its crack growth resistance under conditions of impact or rapid loading.

injection moulded Often refers to the shaping or forming of a *thermoplastic* material by injection through a suitable die at elevated temperature.

Glossary

interstitial A non-metallic atom in an alloy or compound small enough to occupy sites in the lattice between the metal atoms. In a dilute solution of interstitials in a metal, such as carbon in iron, interstitial atoms can migrate (diffuse) without the need for *vacancies.*

ion implantation A sophisticated process in which ions are impacted into the surface of samples causing modifications in properties of the surface layers. Since the process occurs at low temperatures the implanted ions cause extreme distortion in the lattice structure of the substrate material, greatly affecting its properties.

lattice The geometrically perfect network structure on which crystalline materials are based.

martensite, martensitic The phase resulting from a diffusionless (martensitic) phase transformation. In steel martensite is a solid solution of carbon in iron.

matrix The dominating lattice structure of a given phase in a material.

modulus A constant of the elastic strength of a material. For example, Young's modulus (E) refers to the constant of elasticity, or stiffness of a material.

order This usually refers to the structure of an alloy of *stoichiometric* composition. For example, compounds such as aluminium oxide (Al_2O_3) or titanium nitride (TiN) are perfectly *ordered*, each component species forming its own lattice structure, and with the combination forming a *superlattice* structure. It should be noted, however, that in the present text *order is* sometimes used to describe the symmetrical crystalline state of a metal, as opposed to an unordered amorphous state .

phase A microstructural constituent of one given lattice structure or type. A material may contain from one to several coexisting phases.

physical vapour deposition A process for depositing extremely thin layers of material onto a substrate through a vapour or gaseous phase passing over it.

plasma spraying A process for spraying materials onto a substrate or surface in which the materials, in powder form, are heated in a hot plasma created by an electric arc. The technique is

Tomorrow's Materials

particularly useful for laying down surface coatings of ceramic materials.

plastic deformation An irreversible change of shape of a material, brought about in crystalline materials by the creation and movement of large numbers of dislocations.

polycrystalline This refers to a crystalline material containing several (usually very many) grains.

precipitate A minor phase or particle that has precipitated by the movement and coalescence of individual atoms from solid solution during heat treatment.

precipitation-hardened This refers to a material in which the precipitation reaction has been so finely dispersed that the material becomes hardened. The small particles provide a barrier to dislocation movement during *plastic deformation.*

quenching This is a term, commonly used in the heat treatment of metals, in which a certain microstructure or property is obtained by rapidly cooling the sample from some high temperature condition. Blacksmiths of old have used this technique, for example, when hardening steel by quenching it from a glowing-red temperature to oil or salt water, and forming martensite (a solution of carbon and iron).

recycle This is a function of growing importance in today's energy and ecologically conscious world. It is much cheaper to remelt scrap metal than produce new metal from the ore. Recycling of materials today, however, is beginning to involve most materials, including all metals, paper, plastics and glass.

refractory This is a material that can withstand high temperatures. Traditionally it refers to ceramic bricks used to line furnaces.

resources This refers to natural products of the Earth that are utilised by man, such as the minerals and forest.

segregation In this process alloying elements diffuse out of *solid solution* and collect at defects in the material such as grain boundaries.

SIALON This is a commercially successful ceramic alloy, comprising of silicon (Si), aluminium (Al), oxygen (O) and nitrogen (N). In its alloy form it thus becomes: (Si)(Al)ON.

Glossary

sintering The fusion of metallic or ceramic powders into a dense whole when held at elevated temperature. The process occurs in the solid state by interparticle diffusion.

solid solution Very few practical materials are pure in form or consist of only one element. If one element or more are dissolved completely in a base material in the solid state, then the whole is referred to as a solid solution.

spectroscopy When a focussed beam of X-rays or electrons interacts with a material, the emitted X-rays or electrons become highly characteristic of the material through which they have passed. These emissions produce well defined spectra and provide a "finger print" of the material being studied.

stress corrosion A combination of mechanical stress and corrosion which causes an accelerated form of corrosion in the metal.

substitutional This refers to atoms of an alloying addition to a metal which occupy, or substitute, positions or atoms of the matrix material.

superalloys Multiphase alloys, usually cobalt- or nickel-based, which are very heat resistant.

superconductor A material which has practically zero resistivity. The phenomenon, which is thought to be associated with a concerted movement of coupled electrons, occurs in only a few materials and only below some critical temperature.

superplastic Most metals are plastic in the sense that they can be plastically deformed up to a reduction in area at fracture of some 20 %. Superplastic metals can be deformed several 100 % before fracturing. The mechanism by which this occurs does not involve dislocations or glide planes, but occurs instead by a rotation of grains by diffusion at grain boundaries.

thermoplastic A polymer that can be plastically shaped or worked at elevated temperature, hence thermo-plastic. Polyester and PVC are examples.

thermoset A polymer that cannot be plastically shaped or worked at elevated temperature. It is thus thermo-set (in shape). Epoxy is an example.

toughness A property of a material describing its degree of resistance to crack growth or fracture.

Tomorrow's Materials

transformation The transition from one crystalline phase to another. A phase transformation can be brought about either by a diffusioncontrolled reaction or by a diffusionless (martensitic) shear process. The transformation is brought about essentially by chemical changes in energy of atomic bonding that occur on changing the temperature.

vacancy A 'point defect' in a crystal, an atom missing from the lattice structure. It has the important function of aiding the process of diffusion.

weldability A material has good weldability when it can be joined by fusion welding without problems of cracking or excessive porosity occurring.

Index

A

Acrylic foams 138
Afrika 142, 144
Aircraft 23, 56–59, 62, 63
  – skin 23
  – wings 35, 36
Alumina 79, 80
Aluminium 12, 25, 30–32, 34, 35, 56,
    57, 62, 74, 76, 123, 127–131, 138,
    146, 148, 149, 155, 156
  – lithium alloy 58, 61
Amorphous 5–7, 14, 15, 18, 27, 28,
    68, 71, 99, 104
Ankle (human) 121
Aramid 134
Ash 137
Atom(s) 7, 9, 13, 15, 18–20, 27, 28, 30,
    40, 76, 113
Austenite 20, 47
Australia 115, 148
Automobile(s) 21, 23, 84, 155, 157,
    160

B

Badminton 123
Baseball 139
Bats 139
Beryllium 32, 56
Bicycle(s) 57, 69, 70, 127–128, 154
Bismuth 114, 115
Biodegradable 2, 158–160, 162
Bioglass 88
Board(s) (surf) 128–134
Boat(s) 21, 128–134
Bone 35, 88, 121
Boron 34, 102, 107
Brass 31

Brazil 144
Brick 31, 34
Bridge(s) 38, 46
Britain 159
Brittle 14, 32
Bus(es) 154

C

Cable(s) 97, 115, 116, 148
Cadmium 147
Can(s) 52, 56, 57
Canada 148, 159
Canoe(s) 133, 139
Car(s) 30, 56, 69, 127, 154
Carbide(s) 23, 46, 48, 127, 147
Carbon 20, 23, 33, 34, 62, 76, 127, 130,
    138, 143, 146, 153
Cast (alloys) 58
Casting(s) 58
Cedar 132
Cell(s)
  – face centred cubic 20
  – solar 91
– wall(s) 24, 27
Cellular 2, 4, 5, 24–26, 35–37, 88, 125,
    136, 154, 162
Cement 21
Ceramic(s) 2, 4, 11, 12, 14, 20, 32, 33,
    71, 76–91, 111, 126, 137, 138, 148,
    150, 162
Charpy (impact test) 38
Chip(s) 2, 36, 100, 102, 103–106
Chromium 76, 127, 146, 148
Coal 149, 153
Cobalt 147, 149
Combustion engine(s) 143
Composite(s) 2, 4, 5, 21–23, 28, 31,
    33–35, 54, 119, 124, 128, 133, 140

Tomorrow's Materials

Concrete 2, 21, 22, 31, 32, 34, 35, 46, 51–54, 143, 155
Conductivity 15, 24, 104, 105
Conductor(s) 14, 55
Co-polymer(s) 67, 68, 132
Copper 13, 30, 58, 97, 113–116, 147, 148, 155, 156
Cork 25, 26
Corrosion 49, 56, 57, 68, 69, 127
Cost effective 56
Covalent 13, 92
Crack (opening displacement test) 38
Cracking 16, 30
Creep 30, 68, 81
Crystal(s) 7, 8, 9, 11, 15–18, 20
Crystalline 5, 6, 11, 15, 16, 18–21, 27, 67, 73, 99
Crystallisation 21, 28, 29, 91

D

Deep drawing 47
Deforestation 142, 144
Density 33–35, 123
Design 17, 35, 56
Designer 4, 35, 36, 88
Diamond(s) 13, 32
Diesel engine(s) 84
Dinghy 128, 132
Dislocation(s) 7–10, 17, 27, 47, 57
Door(s) 25, 36
Dragonflies 4
Drill(s) 148
Ductility 12, 47, 57
Duralumin 58

E

Ecosphere 142
EEC 148, 152
E-glass 23, 132
Elastic (modulus) 125
Elastomer 67, 68

Electrical
 – conductor 15
 – resistance 38, 92, 105
 – wires 30
Electromagnetic 97
Electron(s) 13, 14, 101, 102, 114
 – pairs 113, 114
Electron (microscope) 7, 39, 44
Elements 146
Energy 19, 20, 154
Entropy 19
Environmental 142, 153, 160
Epoxy 32, 33, 125, 127, 129, 132, 133, 138
Erosion 88, 143
ET (ethylenedithio tetrathiafulvalene) 2
Ethafoam 133
Europe 142
EVA (ethyl vinyl acetate) 123
Extrusion 105

F

Ferrite 20, 48
Ferrites 107
Fibre(s) 23, 57, 127, 133, 162
Fibre composite(s) 64, 134, 138, 140
Fibre optic(s) 36, 92–99, 105
Fibre reinforced 2, 59, 60, 128
Fluoride(s) 95, 99
Flux pinning 115
Foot (human) 119–121
Fossil fuel 143, 149, 153
Fracture 24, 28, 52
 – toughness 31, 79
Friction 130
Fuel consumption 2, 154
Fusion welding 50

G

Gallium arsenside 2, 101–105

Index

Gas 143, 149, 153
Gas turbine(s) 36
Germanium 109
Gold 32, 147, 149, 156
Glass 7, 9, 14, 18, 25, 28, 31, 32, 34, 91, 93
 – fibre(s) 21, 23, 24, 52, 54, 124, 125, 130–132
Glue 2, 4
Grain
 – boundaries 16, 44, 45, 47
 – size 17, 48, 79
Graphite (fibres) 125, 127, 129, 133, 135, 139

H

Hardness 20
 – testing 38
Heat (affected zone) 50, 58
Heat
 – resistant 30
 – treatment 38, 71, 74
Helicopter(s) 15, 58, 61
Hickory 137, 139
Honeycomb 4, 25, 35, 62, 130, 137
Hot isostatic pressing (HIP) 77
Hot pressing 76
Hydrogen 13

I

Ice 13, 15, 32
Ice hockey club(s) 139
Impurity(ies) 8, 17, 31
Inclusion(s) 8, 38
Intermetalic(s) 75
Ion(s) 40, 101
Ion implantation 91
Iron 13, 20, 46, 107, 146, 148, 149, 155, 156
Italy 159

J

Japan 115, 143, 148, 152

K

Kayak 128, 133
Kevlar 32, 126, 128, 132, 134, 139, 140
Kinetic(s) 19
Kitchen utensils 56, 83, 84

L

Lanthanum 109, 114
Laser 71, 72, 79, 104
Lead 148, 149, 156
Leather 119
Leaves 4, 5, 59
Levitation 117
Liquid (state) 6, 11
 – air 111
Lithium 58, 61, 127

M

Machine tool(s) 71, 76, 86
Madagascar 144
Magnesium 31, 34, 56, 57, 127, 128
Magnetic 14, 107–109, 115, 117
Magnet(s) 107, 108, 112
Mahogany 117
Manganese 75, 76, 127, 145–148
Martensite(ic) 50, 72, 79, 80
Meissner effect 117
Melt 18, 74
 – spinnning 73
Mercury 111, 155
Metal(s) 2, 4, 7, 10, 12–14, 20, 25, 33, 100, 148, 154
Microscope(s) 38, 39
Molybdenum 75, 127, 147–149

173

N

Nickel 147–149
Niobium 111, 147
Nitrides 47, 48, 147
Nitrogen 2, 47, 76, 114
Nitrous oxide 143
Nomex 130
North America 142
Nuclear energy 149, 153
Nylon 31, 126

O

Oar(s) 134, 139
Oil 143, 149, 153, 154
Oil platform(s) 38, 44
Optical materials 91–99, 162
Optics 94
Oxide 2, 49, 71, 109, 114, 115, 147
Oxygen 9, 11, 76, 115

P

Paddle(s) 134, 139
Paper 154, 155, 162
Partial dislocation 9
Peek (polyether – ketone) 62, 65, 69
Peru 144
Peroleum 153
Phase (transition or transformation)
    19, 47, 49
Phonon(s) 113, 114
Phosphate 149
Phosphorus 101, 102
Plastic(s) 2, 3, 106, 154, 158, 159, 162
Plastic deformation 7, 8, 11, 16
Plasticity 14
Plama spraying 72
Platinum 32, 147
Pollution 142, 160
Polycrystal (line) 16, 17, 77, 162
Polyacetylene 106

Polyamide 123
Polydiacetylene 108
Polyester 59, 124
Polyethylene 23, 31, 32, 65, 133, 134
Polymer(s) 3, 4, 12, 15, 18, 33, 51–53,
    59, 62, 64, 65, 67, 69, 70, 105, 108,
    119, 125, 127, 134, 138
 – alloys 65, 158, 159
 – chains 6, 28, 29, 159
Polypropylene 65
Polystyrene 65, 67, 70, 130, 132, 159
Polythene 61, 134
Polyurethane (PU) 32, 69, 137
Polyvinyl chloride (PVC) 31, 63, 132
Potassium 91, 92
Pottery 78, 79
Powder
 – metallurgy 71
 – particle(s) 73, 74, 76, 77, 162
Power transmission 116
Precipitate 47, 49, 83
Prosthetic device(s) 87
PSZ (partially stabilized zirconia) 79,
    83–85
Punt 128

Q

Quench 47

R

Racket(s) see racquet
Racquet(s) 123–126
Railway
 – carriages 56
 – lines 117
 – trains 154
Rainforest 143–145
Rapid solidification 73–76
Recycling 3, 30, 154–158, 162
Refractory 81, 86

Index

Resources 142, 161
Roofing 56, 154
Rowing 128
Rubber 14, 25, 28, 31, 33

S

Salt 13
Sandwich panels 4
Scanning electron microscope (SEM) 38–40
Scrap 30, 155, 157
Seashell(s) 4
Semconductor(s) 15, 36, 100, 101, 103, 105, 106
Ship(s) 38, 46
SIALON 81, 147
Silica 9, 18, 32, 90, 95, 98
Silicon 9, 11, 12, 41, 78, 91, 99, 100–106, 126, 162
Silver 33, 147, 149, 156
Sintering 79
Ski(s) 21, 36, 59, 134–139
Slip bands 28
Solar cell(s) 91
Solar energy 99, 162
Solidification 15, 18, 73
South Africa 148
South America 142
South East Asia 142
Soviet Union 148
Spacecraft 58, 59
Space shuttle 88, 90
Specific (property) 34, 57, 128
Spectroscopy 38, 40
Spider(s) 4
Sport(s) 119–139
Sports shoes 119–123
Spruce 132
Squash (racquets) 123, 125
Steel 20, 30–32, 34, 35, 44, 46, 48–50, 74, 75, 100, 127, 138, 157, 162
 – stainless 50

Stiffness 23, 24, 34, 36, 125, 128, 138
Strength 17, 24, 32, 34, 128, 138
Stress
 – corrosion 58
 – flow 4
Strontium 109
Submarine(s) 38
Sulphur 149
Superalloy(s) 30
Superconducting 2, 14
Superconductor(s) 101, 103, 109–117, 162
Superplastic 15
Surface treatment 50, 71–73
Surfboard(s) 132, 133
Sweden 115, 159
·Synthetic 18, 53, 125

T

Telephone 94, 98
Tennis 123, 125
Tensile specimen 11
Textile(s) 154
Texture 57
Thallium 114, 115
Thermal conductivity 105, 106
Thermal expansion 85
Thermodynamic(s) 18, 19
Thermoplastic 22, 23, 67, 70
Tin 147, 149, 156
Titanium 31, 34, 50, 56, 57, 127, 146
Tool(s) 74, 75, 148
Toughness 12, 129, 137, 138
Transformation (phase) 21
Transformation toughening 79
Transmission electron microscope (TEM) 40, 44, 45
Tungsten 32, 75, 147–149
Turbine 36, 84, 86, 87

Tomorrow's Materials

U

U.S.A. 148, 150, 161
U.S.S.R. 148, 151

V

Vacancies 116
Valence electrons 13, 113
Vanadium 127, 147
Vaulting pole 59

W

wavelength spectroscopy 38
Wear-resistant 24, 46, 71, 74, 86
Weld 50
Weldability 46, 58, 127
Welding 49, 58
Wildlife 142
Willow 139
Wing(s) 4, 35, 59

Wood 25, 26, 27, 31, 32, 34, 35, 52, 54, 55, 62, 123, 129, 132, 134, 137, 139, 143, 144, 149, 154, 162

X

x-ray(s) 38, 40

Y

Yacht(s) 128
Yield strength 17, 32, 47
Young's modulus (E) 23, 34, 35
Yttrium 114, 115

Z

Zaire 144
Zinc 32, 56, 57, 147, 149, 156
Zirconia 79, 80, 83
Zirconium 71, 147